Textbooks in Telecommunication Engineering

Series Editor

Tarek S. El-Bawab, Professor and Dean, School of Engineering, American
University of Nigeria, Yola, Nigeria

Dr. Tarek S. El-Bawab, who spearheaded the movement to gain accreditation for the telecommunications major is the series editor for Textbooks in Telecommunications. Please contact him at telbawab@ieee.org if you have interest in contributing to this series.

The Textbooks in Telecommunications Series:

Telecommunications have evolved to embrace almost all aspects of our everyday life, including education, research, health care, business, banking, entertainment, space, remote sensing, meteorology, defense, homeland security, and social media, among others. With such progress in Telecom, it became evident that specialized telecommunication engineering education programs are necessary to accelerate the pace of advancement in this field. These programs will focus on network science and engineering; have curricula, labs, and textbooks of their own; and should prepare future engineers and researchers for several emerging challenges.

The IEEE Communications Society's Telecommunication Engineering Education (TEE) movement, led by Tarek S. El-Bawab, resulted in recognition of this field by the Accreditation Board for Engineering and Technology (ABET), November 1, 2014. The Springer's Series Textbooks in Telecommunication Engineering capitalizes on this milestone, and aims at designing, developing, and promoting high-quality textbooks to fulfill the teaching and research needs of this discipline, and those of related university curricula. The goal is to do so at both the undergraduate and graduate levels, and globally. The new series will supplement today's literature with modern and innovative telecommunication engineering textbooks and will make inroads in areas of network science and engineering where textbooks have been largely missing. The series aims at producing high-quality volumes featuring interactive content; innovative presentation media; classroom materials for students and professors; and dedicated websites.

Book proposals are solicited in all topics of telecommunication engineering including, but not limited to: network architecture and protocols; traffic engineering; telecommunication signaling and control; network availability, reliability, protection, and restoration; network management; network security; network design, measurements, and modeling; broadband access; MSO/cable networks; VoIP and IPTV; transmission media and systems; switching and routing (from legacy to next-generation paradigms); telecommunication software; wireless communication systems; wireless, cellular and personal networks; satellite and space communications and networks; optical communications and networks; free-space optical communications; cognitive communications and networks; green communications and networks; heterogeneous networks; dynamic networks; storage networks; ad hoc and sensor networks; social networks; software defined networks; interactive and multimedia communications and networks; network applications and services; e-health; e-business; big data; Internet of things; telecom economics and business; telecom regulation and standardization; and telecommunication labs of all kinds. Proposals of interest should suggest textbooks that can be used to design university courses, either in full or in part. They should focus on recent advances in the field while capturing legacy principles that are necessary for students to understand the bases of the discipline and appreciate its evolution trends. Books in this series will provide high-quality illustrations, examples, problems and case studies.

For further information, please contact: Dr. Tarek S. El-Bawab, Series Editor, Professor and Dean of Engineering, American University of Nigeria, telbawab@ieee.org; or Mary James, Senior Editor, Springer, mary.james@springer.com

Franco Callegati • Walter Cerroni • Carla Raffaelli

Traffic Engineering

A Practical Approach

 Springer

Franco Callegati
Department of Computer Science and
Engineering (DISI) Cesena campus
Alma Mater Studiorum – Università di
Bologna
Cesena, Italy

Walter Cerroni
Department of Electrical, Electronic and
Information Engineering "Guglielmo
Marconi" (DEI) Cesena campus
Alma Mater Studiorum - Università di
Bologna
Cesena, Italy

Carla Raffaelli
Department of Electrical, Electronic and
Information Engineering "Guglielmo
Marconi" (DEI)
Alma Mater Studiorum - Università di
Bologna
Bologna, Italy

ISSN 2524-4345   ISSN 2524-4353   (electronic)
Textbooks in Telecommunication Engineering
ISBN 978-3-031-09588-7   ISBN 978-3-031-09589-4   (eBook)
https://doi.org/10.1007/978-3-031-09589-4

This Springer imprint is published by the registered company Springer Nature Switzerland AG
The registered company address is: Gewerbestrasse 11, 6330 Cham, Switzerland

This book is dedicated to our past and future students, the true motivation behind this effort, and to Prof. Giorgio Corazza, who stimulated us to the applications of teletraffic engineering as a mentor and friend.

Foreword

The science of queuing and traffic theory dates back to the beginning of the twentieth century and has many applications. The concept of queuing and being served reflects a human behavior. Waiting for a service is indeed part of our daily life. Waiting to be served in a shop, boarding an airplane, waiting for delayed trains, or queuing with your car during the rush hour, or even parents in a busy family, transporting children to activities. In our modern electronic world, queuing has emerged in the form of, for example, software applications awaiting service from the operative system and waiting for access transferring information through an electronic network.

The topic of this book addresses the modern world science of queuing, describing everything from handling access to people (clients) doing voice calls through a traditional central office (the server)-based telephone system to handling data-packets in IP-based networks and, for example, access and data-transfer in 5G mobile networks. Dimensioning a network properly is crucial. Over-dimensioning ensures high performance, but at the cost of expensive network components, driving up the cost of the service production. Under-dimensioning on the other hand typically results in very poor performance, such as audio speech breaking up and frozen video pictures. The practical application of queuing theory is therefore mandatory for proper network design.

Often queuing theory is associated with complex theoretical mathematical models that are hard to grasp and the practical application even more difficult to perform. This book, on the other hand, gives a practical guide to queuing theory, laying the basics for practical network design. The book introduces and explains theory of queuing and queue modelling, always along with practical examples. Engineering and dimensioning of networks, based on both circuit and packet switching, are explained with emphasis on practical applications with up-to-date examples on, for example, IoT. This book is therefore suitable for a broad audience engaged in tele- and data-com networking: everyone, from first-year telecom students to experienced hands-on telecom engineers in need of a practical guide to network dimensioning.

The book is a very welcomed contribution to clarifying and explaining the practical use of a complicated theoretical subject.

Strategic Competence and Research Manager, Tampnet   Steinar Bjørnstad
Adjunct Associate Professor, NTNU
Senior Research Scientist, Simula OsloMet

Preface

Studies on telecommunication network traffic engineering date back to the first years of telephony, at the dawn of the twentieth century. It was a highly specialized area of education, providing competencies strictly required for engineers to be employed by telco operators and equipment vendors.

The deregulation of the telecommunication market and the advent of the Internet, which characterized the last two decades of the twentieth century, modified the telecommunication landscape leading to operators shifting their interest from network fine-tuning and optimization to more aggressive marketing and cost reduction strategies. At the same time, networks evolved from large, monolithic systems managed under the umbrella of public governments to pervasive infrastructures operated by a plethora of entities, ranging from public operators to private companies and individual citizens.

The arena of network engineering has changed a lot in these years, yet some of the basic problems to be solved are still there, and even became more relevant than before. Networks represent a critical part of our daily life, as well as influence a plethora of social processes. Correct network dimensioning and planning are key to providing acceptable levels of quality of service, to fostering sustainable growth, and to enabling effective maintenance and reliability. Proper provisioning and procurement of devices, development of strategies to minimize bottlenecks, and establishment of service differentiation policies are only a few examples of the benefits of accurate network planning.

We believe that queuing theory applied to telecommunication network traffic engineering, also known as *teletraffic engineering*, has still an important role to play, but as a discipline it must evolve to be more accessible to a wider audience beyond the very specialized professionals. Similarly, traditional teaching instruments must be adapted to be more suited to today's and tomorrow's needs. This book intends to provide a practical, yet theoretically solid approach to understanding and using tools for teletraffic engineering as it applies to today's problems. It covers all the principles of circuit- and packet-switched network engineering, with special attention to current practices and real-life case studies, while still providing the necessary mathematical background.

The overall goal is to provide the reader with in-depth and broad understanding of the main concepts and overall operational framework behind teletraffic problems, and therefore induce the capability to select the most suitable and effective method to solve network traffic engineering problems they may face in real life. The reader will learn to pick and choose from a spectrum of tools, ranging from the simplest mathematical formulas to more sophisticated models and case studies.

Cesena, Italy   Franco Callegati
Cesena, Italy   Walter Cerroni
Bologna, Italy   Carla Raffaelli

Contents

1 Introduction to Teletraffic Engineering 1
  1.1 What Is Traffic Engineering? Some Basic Concepts and
       Definitions 1
       1.1.1 The Definition of Traffic 1
  1.2 An Important and Very General Rule of Teletraffic
       Systems: Little's Theorem 3
  1.3 A More Detailed Model of the Teletraffic System: A
       Queuing System 4
       1.3.1 Naming a Queuing System 6
       1.3.2 Little's Theorem for Queuing Systems 8
  Exercises 10

2 An Introduction to Queuing System Modeling 13
  2.1 Introduction 13
  2.2 Modeling Service Requests 14
       2.2.1 The Poisson Process 15
  2.3 Modeling Service Time 23
       2.3.1 Exponential Service Time 24
       2.3.2 Deterministic Service Time 25
       2.3.3 Uniform Service Time 25
       2.3.4 Erlang Service Time 26
       2.3.5 Pareto Service Time 27
  2.4 Link the Time of Arrivals with the Service Time 29
  2.5 Residual Service Time 30
       2.5.1 The Residual Exponential Service Time and Its
              Memoryless Property 32
       2.5.2 The Residual Deterministic Service Time 33
       2.5.3 The Residual Uniform Service Time 34
  2.6 Examples and Case Studies 34
       2.6.1 Time Related Tariffs 34
       2.6.2 Deriving the Poisson Formula from Exponential
              Inter-arrivals 39

　　　　2.6.3　Time to Complete Multiple Services 40
　　Exercises 42

**3　Formalizing the Queuing System: State Diagrams and
　Birth–Death Processes** 45
　3.1　Stateful and Time Dependent Systems 45
　3.2　Defining Congestion as a Sample State 46
　　　　3.2.1　The PASTA Property 48
　3.3　Birth–Death Processes 49
　3.4　Queuing Systems, Memoryless Property, and BD Processes 51
　3.5　Examples and Case Studies 52
　　　　3.5.1　The Poisson Process as a Birth-Only Process 52
　　　　3.5.2　Alarm Reporting 53
　　　　3.5.3　Taxis at the Airport 56
　　Exercises 58

4　Engineering Circuit-Switched Networks 65
　4.1　Introduction 65
　4.2　Modeling Circuit Switching Systems Without Waiting Space 66
　　　　4.2.1　Performance Metrics 66
　　　　4.2.2　An Ideal System with Infinite Circuits 67
　　　　4.2.3　The Real System with a Finite Number of Circuits 69
　　　　4.2.4　Utilization of Ordered Servers 74
　　　　4.2.5　Comparing the $\mathcal{M}/\mathcal{M}$ and the $\mathcal{M}/\mathcal{M}/m/0$ Systems 76
　　　　4.2.6　Insensitivity to Service Time Distribution 77
　　　　4.2.7　How Good Is the Erlang $\mathcal{B}$ Model 77
　　　　4.2.8　Examples and Case Studies 78
　4.3　Modeling Circuit Switching Systems with Waiting Space 92
　　　　4.3.1　Performance Metrics 92
　　　　4.3.2　The $\mathcal{M}/\mathcal{M}/m$ System 93
　　　　4.3.3　Waiting Time Probability Distribution for a FIFO Queue ... 98
　　　　4.3.4　Examples and Case Studies 102
　4.4　Multi-Dimensional BD Processes 115
　　　　4.4.1　The Multi-Service Link 119
　　　　4.4.2　Circuit-Switched Networks with Fixed Routing 120
　　　　4.4.3　Examples and Case Studies 123
　　Exercises 133

5　Engineering Packet-Switched Networks 141
　5.1　Introduction 141
　5.2　Single Server Queuing 142
　　　　5.2.1　Performance Metrics 142
　　　　5.2.2　A General Result for Single Server Systems with
　　　　　　　Infinite Waiting Space 143
　5.3　Memoryless Single Sever Queuing Systems 145
　　　　5.3.1　Infinite Queuing Space: The $\mathcal{M}/\mathcal{M}/1$ System 145

	5.3.2	Finite Queuing Space, the $\mathcal{M}/\mathcal{M}/1/L$ System	153
	5.3.3	Examples and Case Studies	156
5.4	A Look at Some More General Cases		169
	5.4.1	Poisson Arrivals May Be Fine, But What About Service Time? The $\mathcal{M}/\mathcal{G}/1$ Queue	169
	5.4.2	Packet Switching and Quality of Service: When One Pipe Does Not Fit All	177
	5.4.3	Examples and Case Studies	184
Exercises...			198

A Brief Introduction to Markov Chains 203
A.1	Discrete Time Markov Chains		203
	A.1.1	Transition and Steady State Probabilities	204
	A.1.2	Irreducible Markov Chains...................................	207
	A.1.3	The Chapman–Kolmogorov Equations	208
	A.1.4	The Markov Chain Behavior as a Function of Time	209
	A.1.5	Time Spent in a Given State	210
A.2	Continuous Time Markov Chain......................................		211
	A.2.1	The Time Spent in a State	214

B The Embedded Markov Chain for the $\mathcal{M}/\mathcal{G}/1$ System 215
| B.1 | Steady State Probabilities of the Number of Customers in the System at Departure Times .... | | 216 |
| B.2 | Steady State Probabilities at Generic Time Instants .... | | 219 |

Index... 221

Chapter 1
Introduction to Teletraffic Engineering

Abstract This chapter will introduce the reader to teletraffic engineering. The concept of *traffic* will be formally defined, together with the *teletraffic system*. Then the basic mathematical concepts of arrival rate and sojourn time are defined. These quantities are fundamental to proceed with a quantitative dissertation and are linked to traffic by means of the Little's Law (or Theorem), which is a fundamental result enunciated here and widely used in the book. In the second part of the chapter, the teletraffic system of interest for the book is better specified by defining the concept of *queuing system* and related characteristics, thus setting up the basis for what is presented in the remainder of the book.

Keywords Teletraffic engineering · Erlang · Little · Little's theorem · Queuing system · Queuing performance · Average customer arrival rate · Average customer service time · Kendall's notation · Scheduling · FIFO · LIFO · Offered traffic · Lost traffic · Throughput · Utilization

1.1 What Is Traffic Engineering? Some Basic Concepts and Definitions

Teletraffic engineering is a topic born after the deployment of the first large telecommunication networks at the very beginning of the XX century. It is mostly based on the pioneering studies of the Danish mathematician Agner Krarup Erlang, who provided several important contributions in this field. For this reason, Erlang's name will frequently come out along the book. This introductory chapter will provide the basic definitions and concepts that will be used in the following.

1.1.1 The Definition of Traffic

At first, let us consider a generic system called *teletraffic system*. It is a black box into which *customers* (or *users*) wish to enter to accomplish some form of task or

© Springer Nature Switzerland AG 2023
F. Callegati et al., *Traffic Engineering*, Textbooks in Telecommunication Engineering, https://doi.org/10.1007/978-3-031-09589-4_1

service (for the time being, the specific purpose of the task or service is not important and may be whatever). Once entered, the customers stay in the system for some time and, after having accomplished the task (service expired), they leave the system.

Some basic but very relevant assumptions are:

- the service is atomic, meaning that every customer has its own needs that have to be serviced separately from the needs of other customers (even though the service of different customers may have similar properties);
- users do not "appear" or "disappear" by magic: if a user is in the system he/she came from the outside at some time and, once entering the system, he/she will also leave after some time.

These assumptions imply that the observation of the teletraffic system from the outside provides all the information required to understand how many users are currently in the system itself. This is of particular importance given that, in general, the focus of the analysis of a teletraffic system is on the number of customers in the system.

> $k(t)$ is the number of customers in the system at time t and is called the *traffic in the system at time t*.

Given that the behavior of the users is usually random, in general $k(t)$ is a random process. For this reason, teletraffic problems are typically studied by means of probability theory.

We can calculate the average traffic in a time interval T as:

$$I(T) = \frac{\int_T k(t)dt}{T}$$

> When it exists, the limit
>
> $$A = \lim_{t \to \infty} I(t) = \langle k(t) \rangle \tag{1.1}$$
>
> defines the *average traffic intensity*. If the limit does not exist, the system does not have a stable behavior and cannot be studied with the analysis presented here.

The average traffic intensity A has no physical dimension but is conventionally measured in *Erlangs*.[1] Therefore, it is common to say that "A is equal to 10 Erlangs" or "$A = 10$ E" rather than "A is equal to 10."

---

[1] In view of the ground-breaking work of A. K. Erlang, the International Consultative Committee on Telephones and Telegraphs (CCITT) honored him in 1946 by adopting the name "Erlang" for the basic unit of telephone traffic.

If $k(t)$ is an ergodic process, A exists and is equal to the mean of the process at a generic time t. This means that the study of the statistics of the system at a generic time instant can predict the statistics of one specific behavior of the system in time.

Therefore, it is desirable to deal with ergodic traffic. Since the real users behavior is not ergodic in general, the usual way to proceed is:

- select a time period that is considered significant from the system engineering point of view and during which the traffic can be considered ergodic;
- analyze the random behavior of the traffic in such period;
- assume the traffic is ergodic with such behavior all over the lifetime of the system.

1.2   An Important and Very General Rule of Teletraffic Systems: Little's Theorem

Let us define:

$a(t)$ as the total number of arrivals to the system at time t,
$d(t)$ as the total number of departures from the system at time t,
δ_n as the time spent in the system by the n-th customer.

It is easy to understand that:

$$k(t) = a(t) - d(t) \tag{1.2}$$

The average arrival rate of customers to a traffic system is given by the following limit:

$$\lambda = \lim_{t \to \infty} \frac{a(t)}{t} \tag{1.3}$$

The limit may or may not return a finite result. In the former case the average arrival rate exists, in the latter case the average arrival rate cannot be defined.

The average time spent in the system by the users is given by the following limit:

$$\bar{\delta} = \lim_{t \to \infty} \frac{\sum_{n=1}^{d(t)} \delta_n}{d(t)} \tag{1.4}$$

The limit may or may not return a finite result. In the former case the average time spent in the system exists, in the latter case the average time spent in the system cannot be defined.

Little's Theorem links A, λ, and $\bar{\delta}$. If all these quantities exist, Little's Theorem says that:

$$A = \lambda \bar{\delta} \tag{1.5}$$

In simple words, if the average customer arrival rate and the average time spent in the system by a customer exist and have finite value, the average traffic in the system is simply given by their product.

1.3 A More Detailed Model of the Teletraffic System: A Queuing System

The *queuing system* is a kind of teletraffic system with some specific characteristics. The system *users* require some specific *service* and are therefore usually considered *customers* of the system. The service is provided by *servers* that exist in the teletraffic system.

Unless we explicitly say otherwise, in this book the service is considered one unique thing, meaning that when a server takes care of serving a customer, the service will be provided just by that server and will be completely performed as one single task without interruptions. Moreover, if the number of servers is finite, it is possible that customers want to enter the queuing system when all servers are busy serving previous customers. In this case, the customers may be allowed to enter the system to wait for a server to become free and the system has to be equipped with some sort of *waiting space*, which will be called a *queue*. It may happen that the waiting space is not present, or it has a finite size and is already full of customers. In that case, a request to enter the system may be *rejected*, i.e., the customer is not allowed to enter the queuing system. Figure 1.1 shows the generic scheme of a queuing system, with several servers (little circles) and a queue (the stairs like rectangle) with waiting spaces.

Some other nontrivial assumptions that will be used in the following are:

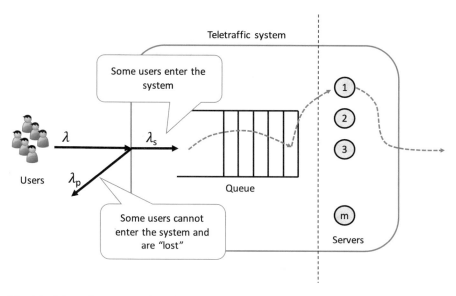

Fig. 1.1 Schematic representation of a queuing system

- the time required to transfer a customer from the queue to a server is considered negligible and is not taken into account in the calculations that follows or can be considered part of the service time, with a negligible influence on its statistical behavior;
- a customer, once entered, is not allowed to leave the system without service, i.e., all customers will leave the system after being taken care of by a server;
- the servers have uniform accessibility, meaning that a customer entering the system and finding more than one server free will choose at random by which server to be served; therefore, servers are all identical from the customer perspective;
- the statistics of the service time do not depend on the server but may depend on the customer.

The aforementioned properties are rather intuitive when considered in the perspective of daily life experience. We all have continuous experience of queuing systems in real life, we queue at the post office, at the grocery store, to take the bus or the subway, to enter a sport game, to talk with an operator at a call center, etc. We know that when joining the queue we will have to wait patiently for some time, we know that each of us will hopefully obtain what he/she wanted when joining the queue, and we also know that sometimes waiting in a queue could be very boring because it takes much longer than we wanted to get to the point.

We also know that not all queuing systems are the same: some are rather efficient and usually have a very short or no queue at all; others are not that efficient and always have long waiting queues; in some cases we have just one way to get the service, in other cases we are offered many parallel options (for instance, several clerks), etc. What we intend to do here is to better formalize the description of a queuing system. The goal is to understand what are the features and parameters that matter, whether they are qualities or quantities and, at the end of the day, how they influence the behavior of the queuing system from a qualitative as well as quantitative point of view.

1.3.1 Naming a Queuing System

Let us start with the characterization of a queuing system. For engineering purposes there are a number of characteristics we have to make clear before we can tackle the problem of modeling the system, the most important of which are:

- the statistics of the customer arrival process;
- the statistics of the service process;
- the number m of servers present in the system;
- the size L of the queuing space, if any;
- the size p of the customer population, i.e., how many customers may require service at a specific queuing system;
- the queue scheduling policy, i.e., the strategy the servers will use to retrieve customers from the queue when a service is completed and a new user may be served.

To summarize these characteristics in a more compact form, in the following we will use a well-known representation called *Kendall's notation*.[2] Kendall's notation is a series of symbols separated by slash "/" that in short allows us to provide the aforementioned characteristics all at once, for instance, with a string such as $\mathscr{A}/\mathscr{B}/m/n/p/\mathscr{S}$.

Here m, n, and p are integers while $\mathscr{A}$, $\mathscr{B}$, and $\mathscr{S}$ are symbols taken from a conventional alphabet to provide specific information in a very concise way. The conventional meaning of numbers and symbols is as follows:

$\mathscr{A}$ a conventional symbol to represent the statistics of the arrival process;
$\mathscr{B}$ a conventional symbol to represent the statistics of the service process;
m the number of servers in the system;
n the size of the queuing space;
p the total number of customers;
$\mathscr{S}$ a conventional symbol to represent the scheduling policy.

Regarding the arrival and service process statistics, a simple alphabet that is enough for the topics covered by this book is defined as follows:

---

[2]David George Kendall (1918–2007) was an English statistician and mathematician, teaching at the University of Oxford (1946–1962) and at the University of Cambridge (1962–1985).

$\mathscr{M}$ the statistics refer to a *memoryless*[3] random process;
$\mathscr{D}$ the statistics refer to a *deterministic* process, which is not random at all;
$\mathscr{G}$ the statistics refer to a *general* process about which we do not have detailed information, apart from some statistical moments such as mean and variance.

Regarding the scheduling policies, some typical symbols are:

FIFO First In, First Out (or FCFS First Come, First Served), i.e., customers are served in the same order as they arrive;

LIFO Last In, First Out (or LCLS Last Come, First Served), i.e., the last customer arrived to the system will be the next one to be served;

RC Random Choice, i.e., the next customer to be served is chosen randomly among those that are waiting in the queue;

PRIO scheduling with priorities, i.e., if some kind of customers are waiting in the queue, they will be served before other kinds of customers, no matter the order of arrival.[4]

Finally, there are some additional conventions about Kendall's notation:

- if any of the integers (m, n, p) is missing, a default value equal to infinite is assumed, i.e., infinite servers, infinite queuing space, infinite population of customers;
- if $\mathscr{S}$ is not specified, the default choice is a FIFO scheduling.

> In the remainder of this book, unless stated otherwise, we will assume that the m servers are identical from the customer perspective. This is to say that the customer entering the system will pick one of the servers at random since they will exhibit the same behavior with respect to the service, without providing any form of privilege to one customer or to another.

It is rather clear that infinite values for the queuing space or for the customers population are not possible in general. Nonetheless, this assumption is used very often as a first-step approximation for cases where these numbers are very large.

> In the remainder of this book, unless stated otherwise, we will refer to FIFO systems with an infinite user population.

The FIFO scheduling is very popular and reasonable. It is associated with a general concept of fairness that in many human societies is also assumed to be a "well behaving" attitude. While it is rather intuitive that this is correct at the cashier of the grocery store, it may not be true in general. For instance, in an Emergency Room nobody will complain if a patient with a severe heart attack is treated before

---

[3]The "memoryless" property will be explained in detail in the next chapter.
[4]This is also a concept that will be better treated and explained in the following chapters.

other patients with minor injuries, even if they arrived before and have already been waiting for a while. In this case, the need to schedule the service according to a logic that is not related to the time of arrival is very intuitive. Even more important, the FIFO queue is not necessarily the best policy when customers are not humans, but computers, for instance. In this case the ethical issue of fairness is not really relevant and objectives of efficiency and optimality may lead to the adoption of other scheduling policies, as we will see in the following.

Last but not least, the assumption of an infinite customer population may appear quite unrealistic. Nonetheless, it is widely used and well matching with empirical results in a large variety of cases. The reason is that, especially when dealing with large telecommunications networks, the user population is really large and the amount of traffic generated by a single user is relatively much smaller. Think about the telephone network of a medium-size town: the number of users is likely in the range of tenth if not hundredth of thousands, while any single user uses the telephone rather rarely, probably just for a few minutes per day. Therefore, the number of users actually using the system is in general very low when compared to the whole user population. The consequence is that the statistical behavior of the whole population is almost independent of the number of the users in service, i.e., it behaves as if the total number of users was infinite.

1.3.2 Little's Theorem for Queuing Systems

Some important quantities relevant to the study of queuing systems are defined in the following.

λ_s	average rate of customers entering the queuing system;
λ_p	average rate of customers wishing to enter the queuing system but without success (rejected customers);
$\lambda = \lambda_s + \lambda_p$	average rate of requests to enter the queuing system (including successful and unsuccessful requests);
δ	time spent in the queuing system by a generic customer (*sojourn time*);
η	time spent in the queue of the queuing system by a generic customer (*waiting time*);
ϑ	time spent in service by a generic customer (*service time*);
A_c	average number of customers in the queue;
A_s	average number of customers in service.

Figure 1.2 shows a graphical example of the relationship between waiting time η, service time ϑ, and total time spent in the system δ.

Little's Theorem holds for queuing systems as well and can be written as

$$A = \langle k(t) \rangle = \lambda_s \bar{\delta} \qquad (1.6)$$

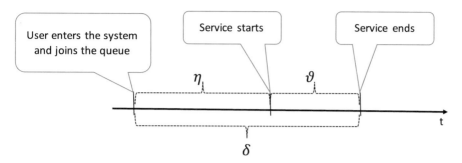

Fig. 1.2 Schematics of the time spent in the system by a customer. The total time in system δ is the sum of the waiting time in the queue η and the service time ϑ

Moreover, if we consider the queue and the set of servers as two different subsystems, Little's Theorem leads to the following equations:

$$A_c = \lambda_s \bar{\eta} \tag{1.7}$$

$$A_s = \lambda_s \bar{\vartheta} \tag{1.8}$$

where $\bar{\eta}$ and $\bar{\vartheta}$ are the average waiting and service times, respectively.

Given the aforementioned assumption that the time to transfer a customer from the queue to the servers is negligible, it is also true that:

$$\bar{\delta} = \bar{\eta} + \bar{\vartheta}$$

and therefore:

$$A = A_c + A_s$$

We will also make extensive use of some additional quantities that are defined as:

$A_0 = \lambda \bar{\vartheta}$	the offered traffic;
$A_s = \lambda_s \bar{\vartheta}$	the throughput (or served traffic);
$A_p = \lambda_p \bar{\vartheta}$	the lost traffic.

Given that $\lambda = \lambda_s + \lambda_p$, it happens that:

$$A_0 = A_s + A_p. \tag{1.9}$$

It is worth noting that, although all the above quantities are defined as "traffic," A_s is the only one actually referring to a real phenomenon, since it measures the average number of busy servers in the queuing system we are considering. Both A_0

and A_p do not refer to any real traffic, they are simply defined as the product of an arrival rate and an average service time. Nonetheless, they are relevant quantities to which we will provide some physical meaning in the following chapters. For instance, A_0 can be seen as the traffic in a system that accepts all service requests and serves them immediately without queuing. Such a system must be equipped with a very large number of servers, as many as the users that may request a service, and therefore is a sort of *ideal system* with an infinite number of servers (see Chap. 4).

Finally, we define the following key performance indicator:

$$\rho = \frac{A_s}{m} \tag{1.10}$$

ρ is called the *server utilization* and is a quantity bound between 0 and 1, since it is impossible that more than m customers are simultaneously served in a system with m servers.

If the queuing system has a random behavior that is ergodic in nature, A_s represents the average number of busy servers in a given instant of time, whereas ρ represents the average percentage of time a server is busy serving customers.

Exercises

1. Assume that $p = 8$ customers arrive to a teletraffic system according to the sequence shown in the table below, where n identifies the customer, t_n is the instant when customer n arrives to the system, δ_n is the time spent in the system by customer n, and d_n is the departure time of customer n (all times are expressed in minutes).

n	t_n	δ_n	d_n
1	1	3	
2	2	5	
3	3	6	
4	6	2	
5	7	3	
6	11	4	
7	12	3	
8	14	2	

(a) Complete the table above by computing the departure time of each customer.
(b) Compute the total number of arrivals $a(t)$ and departures $d(t)$ at time t, for $1 \le t \le 16$.
(c) Draw a graph that shows the behavior of the traffic in the system $k(t)$ as a function of time t, for $1 \le t \le 16$.

2. The average number of phone calls set up at a mobile base station is $\lambda = 200$ calls/min. The average call holding time is $\bar{\vartheta} = 2$ min. Assuming that the base station has enough resources to immediately establish any incoming call without queuing it, find the average number of active calls in the base station, i.e., the traffic in the base station A.

3. Consider a teletraffic system where the average number of active calls throughout the day can be approximated as follows:

$$
A = \begin{cases} 200 \text{ E} & \text{from 8 am to 5 pm} \\ 50 \text{ E} & \text{from 5 pm to 8 pm} \\ 18 \text{ E} & \text{from 8 pm to 8 am} \end{cases}
$$

Assume also that the average sojourn time in the system varies during the day and is measured as follows:

$$
\bar{\delta} = \begin{cases} 8 \text{ min} & \text{from 8 am to 5 pm} \\ 5 \text{ min} & \text{from 5 pm to 8 pm} \\ 6 \text{ min} & \text{from 8 pm to 8 am} \end{cases}
$$

Find the average rate of customers entering the system λ_s in the three time intervals considered above.

4. Consider a medical facility with 10 vaccination points, where a new patient is admitted every minute, according to a rigorous reservation schedule. The building includes a queuing space holding up to 50 patients. When all the seats in the queuing space are occupied, any incoming patient is rescheduled to another time. The total number of patients is so large that can be assumed infinite.

   Whenever a vaccination point becomes available, if there are any patients waiting in the queue of age equal or greater than 60 years, they are given precedence over younger patients and are vaccinated first. The time it takes for any patient to be vaccinated depends on several factors and must be assumed to be random, but it can be considered "memoryless" with an average of 15 min.

   Elaborate on how the medical facility can be described as a queuing system and write the corresponding representation in Kendall's notation. Then compute the traffic offered to the queuing system.

5. The Private Automatic Branch Exchange (PABX) installed in a company is connected to the Public Switched Telephone Network (PSTN) with $m = 20$ lines. The average call arrival rate towards the PSTN is $\lambda = 3$ calls/min and the average call duration is $\bar{\vartheta} = 4$ min. In the assumption that all calls arriving can also be successfully set up (i.e., neglecting the probability for a call to be

rejected)), calculate the throughput of the PABX towards the PSTN and the average utilization of the lines.

6. In a network node, packets arrive at a rate of $\lambda = 800$ packets/s. The packet size is variable: 64 bytes with a probability of 40%, and 1500 bytes with a probability of 60%. A newly arrived packet is immediately served if the output interface on which the packet must be transmitted (the server) is not busy transmitting another packet. Otherwise, the packet is stored in a queue of finite size and served according to a FIFO scheduling policy. Any new arrival that finds a full queue is dropped.

Assuming that the output interface of the node works at 10 Mbits/s, and that 1% of the arriving packets are dropped, compute the server utilization ρ.

7. Customers arriving to an $\mathcal{M}/\mathcal{M}/3$ teletraffic system are served with an average service time $\bar{\vartheta} = 0.1$ s. Compute the maximum value of the customer arrival rate such that the server utilization does not exceed 80%. Assuming that under such condition the average sojourn time is $\bar{\delta} = 0.2$ s, compute the average number of customers waiting in the queue.

Chapter 2
An Introduction to Queuing System Modeling

Abstract To be able to perform quantitative analysis of a queuing system, a related mathematical model is required. In this chapter we will discuss the issue of modeling both customer service requests and customer service time when they are randomly distributed. The chapter will mostly focus on the Poisson model and on its memoryless property. After reading this chapter, the reader will know how to describe a customer arrival process, will learn the mathematical model to describe a Poisson process and how to calculate its relevant quantitative characteristics, and will learn about the memoryless property of an exponentially distributed service time. Finally, some examples of application of these results will be provided.

Keywords Customer arrival time · Customer departure time · Inter-arrival time · Poisson process · Negative exponential probability distribution · Negative exponential probability density function · Merging Poisson processes · Splitting a Poisson process · Residual service time · Memoryless property

2.1 Introduction

As already outlined, for most of the problems relevant to telecommunications it is reasonable to assume that the customers exhibit a random behavior. This is part of our everyday life experience. The use of a mobile phone, of the e-mail, or of the web browser is not linked to some specific time frame or set of permits: we can do as we like and the same is true for the billions of users of such communication systems. Since we are customers of the system requiring some service when performing the aforementioned actions, it is easy to understand that a deterministic description of the customer behavior is not possible.

For this reason it is necessary to make use of probability tools and the theory of stochastic processes to proceed forward. This chapter presents the basic models that will be extensively used in the following to tackle the relevant engineering problems. The first part of the chapter deals with the arrival process and in particular with the Poisson process, which is by far the most used one. The second part will discuss the modeling alternatives for the service time.

2.2   Modeling Service Requests

Figure 2.1 shows a possible way to plot a sample arrival process, a method we will use throughout this book. The crosses in the graph correspond to requests arriving from a set of customers and provide an example of a random behavior. Requests may arrive at any possible instant and may be clustered together or more widely spaced, depending on chance. Therefore, to model such a phenomenon we have to consider stochastic processes. In this book we will just consider one-dimensional processes that are discrete in value. The dimension of the process is time, and time is assumed to be a continuous variable. It is interesting to note that in the example provided there are no overlapping requests, which is an important property of the process, an issue that we will discuss in more detail later.

Now the first question that we have to answer is how to approach a quantitative description and modeling of the arrival process. Indeed there are several possible alternatives, but intuitively it is clear that they have to grasp how the arrivals cluster in time. In this section we will use two alternative modeling approaches that are obviously linked together:

- the probability of the number of arrivals k in a given time interval T that we call $P(k, T)$;
- the statistics of the time between two consecutive arrivals, i.e., the *inter-arrival time*, that we call τ in the following.

These two quantities are shown in Fig. 2.1 with reference to a subset of the arrivals, in the time frame between $t = 10$ and $t = 15$.

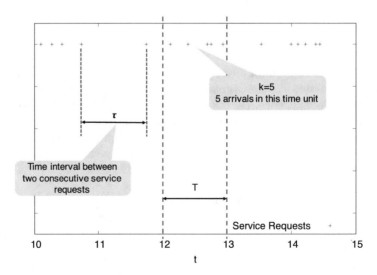

Fig. 2.1 Example of arrivals and quantities that can be used to characterize an arrival process

2.2.1 The Poisson Process

Following the approach mentioned above, in this section we introduce and study the Poisson process. The formulas that will follow can be introduced and demonstrated in different ways. Given the practical focus of this book, we selected an approach that starts from some physical characteristics of the process and then derives the mathematical formulation.

For the time being, let us assume a population of p customers. In a reference time interval T the customers satisfy the following hypotheses:

1. they behave identically, i.e., they request the same type of service and have similar statistical and quantitative behaviors;
2. they request the service independently of one another;
3. over the time interval T the service requests from a single customer are uniformly distributed with probability density function $f_1(t) = \frac{1}{T^\star}$, that is to say that $f_1(t) = \frac{1}{T^\star}$ for $t \in T^\star$ with $T^\star \gg T$.[1]

These are the basic assumptions over which the model will be developed. Their practical applicability and meaning will be further discussed later.

Now let us split T in N sub-intervals of equal size ΔT, choosing $N \gg 1$ such that:

$$\Delta T = \frac{T}{N} \ll T \tag{2.1}$$

Assuming that the N intervals are indexed consecutively with index $s \in [1 : N]$, we are interested in finding the probability of i service requests happening in ΔT.

Let us observe the generic client c. According to hypothesis 3 above, we can calculate the probability $P_a(c, s)$ that c sends a service request in the s-th interval ΔT:

$$P_a(c, s) = \Pr\{\text{one service request in } \Delta T\} = \int_{\Delta T} f_1(t)dt = \frac{\Delta T}{T^\star} \tag{2.2}$$

According to hypotheses 1 and 2, the clients do not change their behavior in time and are all identical. Moreover, all the intervals are of the same size. Therefore, $P_a(c, s)$ does not depend on c and s and can be represented as P_a, that is the probability that a generic customer sends a request in a generic interval ΔT, regardless of which is the specific customer or the specific interval.

Since the customers are p in total, the number of service requests in ΔT can range from 0 to p. As a matter of fact, we can consider this situation equivalent to running p independent experiments during ΔT, each of them with a binary output

---

[1]To simplify the notation used in this section, we represent with $T^\star$ and T both the time intervals and their respective sizes.

that is whether there is an arrival or not. This is like saying that we are tossing a coin p times. The experiments are independent and the probability of success (something happens) is P_a, while the probability of in-success (nothing happens) is $1 - P_a$. Therefore, the binomial distribution with parameters p and P_a can be applied, resulting in the probability of generating i service requests out of p customers in an interval being:

$$P_i = \Pr\{i \text{ requests out of } p \text{ in } \Delta T\} = \binom{p}{i} P_a^i (1 - P_a)^{p-i}$$

$$= \binom{p}{i} \left(\frac{\Delta T}{T^\star}\right)^i \left(1 - \frac{\Delta T}{T^\star}\right)^{p-i} \tag{2.3}$$

If $N \gg p$ and $\Delta T \ll T^\star$, then:

$$P_0 = (1 - P_a)^p \simeq 1 - p P_a = 1 - p \frac{\Delta T}{T^\star} \tag{2.4}$$

$$P_1 = p P_a (1 - P_a)^{p-1} \simeq p P_a = p \frac{\Delta T}{T^\star} \tag{2.5}$$

$$P_i \simeq o(P_a) \quad \forall i > 1 \tag{2.6}$$

In summary, these formulas say that when $\Delta T \ll T^\star$ the probability of having overlapping service requests in the same ΔT is negligible with respect to the probability of having only one service request or none at all.[2]

If we define:[3]

$$\lambda = \frac{p}{T^\star} \tag{2.7}$$

then:

$$P_0 = 1 - \lambda \Delta T \tag{2.8}$$

$$P_1 = \lambda \Delta T \tag{2.9}$$

Now that every ΔT basically represents a random experiment that may have a 0 or 1 result, and given that the various intervals are independent, we can apply the binomial distribution to the N intervals in T, with success probability P_1, and obtain the probability of k service requests in T as:

---

[2] As usual the symbol $o(P_a)$ means that the quantity considered is negligible when compared to P_a.

[3] It is interesting and important to note that the dimension of λ is a frequency (i.e., seconds^{-1}). It will be called *average arrival rate* in the following and is the physical quantity that characterizes the Poisson process.

$$\Pr \{k \text{ requests out of } N \text{ intervals in } T\} = \binom{N}{k} (\lambda \Delta T)^k (1 - \lambda \Delta T)^{N-k} \qquad (2.10)$$

Considering that when $\Delta T \to 0$ obviously $N \to \infty$, it is now possible to compute the following limit:

$$
\begin{aligned}
P(k, T) &= \lim_{N \to \infty} \binom{N}{k} (\lambda \Delta T)^k (1 - \lambda \Delta T)^{N-k} \\
&= \lim_{N \to \infty} \binom{N}{k} \left(\frac{\lambda T}{N}\right)^k \left(1 - \frac{\lambda T}{N}\right)^{N-k} \\
&= \lim_{N \to \infty} \frac{N!}{k!(N-k)!} (\lambda T)^k N^{-k} \left(1 - \frac{\lambda T}{N}\right)^{-k} \left(1 - \frac{\lambda T}{N}\right)^{N} \qquad (2.11) \\
&= \frac{(\lambda T)^k}{k!} \lim_{N \to \infty} \underbrace{\frac{N!}{(N-k)!}}_{N^k} \underbrace{(N - \lambda T)^{-k}}_{N^{-k}} \underbrace{\left(1 - \frac{\lambda T}{N}\right)^{N}}_{e^{-\lambda T}} \\
&= \frac{(\lambda T)^k}{k!} e^{-\lambda T}
\end{aligned}
$$

This result is the so-called *Poisson formula* and is one answer to the original questions of providing a quantitative model to describe the statistics of the arrival process.

The probability of k arrivals in a time interval of size T for a stationary Poisson process depends only on T and on the average process arrival rate λ. It is given by the formula:

$$P(k, T) = \frac{(\lambda T)^k}{k!} e^{-\lambda T} \qquad (2.12)$$

The behavior of formula (2.12) is shown in Fig. 2.2 as a function of λT for different values of k, and in Fig. 2.3 as a function of k for different values of λT.

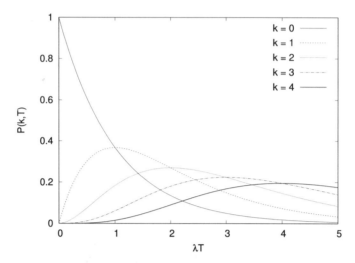

Fig. 2.2 Poisson process: probability of k arrivals in a time interval T as a function of λT, for different values of k

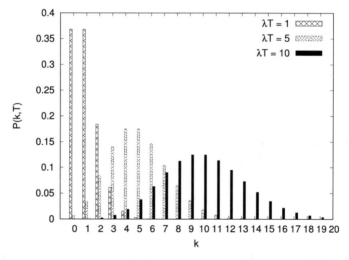

Fig. 2.3 Poisson process: probability of k arrivals in a time interval T as a function of k, for different values of λT

2.2.1.1 Average Arrival Rate

The average number of service requests in T is $\bar{k} = \lambda T$.

It can be easily calculated as follows:

$$\bar{k} = \sum_{k=0}^{+\infty} k P(k, T) = \sum_{k=1}^{+\infty} \frac{(\lambda T)^k}{(k-1)!} e^{-\lambda T} = (\lambda T) e^{-\lambda T} \underbrace{\sum_{k=1}^{+\infty} \frac{(\lambda T)^{k-1}}{(k-1)!}}_{e^{\lambda T}} = \lambda T$$

(2.13)

> The variance of the number of service requests in T is $var(k) = \lambda T$.

It can be easily calculated as follows:

$$var(k) = \sum_{k=0}^{\infty} (k - \bar{k})^2 P(k, T) = \lambda T$$

(2.14)

> In a Poisson process the average number of requests and the variance of the number of requests in T are equal, i.e., $\bar{k} = var(k)$.

It is interesting to notice that the Poisson process has this specific characteristic of showing a variance exactly equal to the mean value.

2.2.1.2 Inter-arrival Time

In the beginning we mentioned that the statistics of the inter-arrival time τ could be another method to grasp the process behavior with mathematical formulas. Such statistics can be easily derived from the probability $P(k, T)$, now that we know it. Say we have the n-th service request at time t_0 and the $n + 1$-th at time t_1. In this specific case the inter-arrival time is $\tau_{n,n+1} = t_1 - t_0$. We note that $\tau_{n,n+1} \geq t$ if and only if there are 0 request arrivals between t_0 and $t_0 + t$. Therefore, considering a generic inter-arrival time τ, we can say that:

$$\Pr\{\tau \leq t\} = 1 - \Pr\{\tau \geq t\} = 1 - P(0, t)$$

(2.15)

We can calculate the probability distribution and density functions of τ, respectively, as:

$$F_\tau(t) = \Pr\{\tau \leq t\} = 1 - e^{-\lambda t} \quad t \geq 0$$

(2.16)

$$f_\tau(t) = \begin{cases} 0 & t < 0 \\ \lambda e^{-\lambda t} & t \geq 0 \end{cases}$$

(2.17)

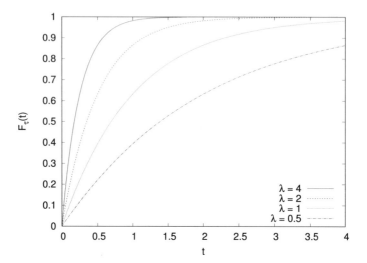

Fig. 2.4 The negative exponential probability distribution function plotted for some values of λ

These distribution and density functions are called *negative exponential* or simply exponential.

> The probability distribution describing inter-arrival time in a Poisson process is the exponential distribution.

The average arrival rate λ is the constant that determines the actual value of the distribution, as well as the probability of k arrivals in T. The exponential distribution has a very important property that is crucial to make the Poisson process so useful in teletraffic theory, namely the *memoryless property*. We are not going to discuss such property here but will devote to it a whole section in the remainder of the chapter. The behaviors of the exponential probability distribution and density functions for different values of λ are shown in Figs. 2.4 and 2.5, respectively.

Calculating average and variance of τ is now rather straightforward:

$$\bar{\tau} = \int\limits_0^{+\infty} t f_\tau(t)dt = \int\limits_0^{+\infty} t\lambda e^{-\lambda t}dt = \frac{1}{\lambda} \tag{2.18}$$

$$\sigma_\tau^2 = \frac{1}{\lambda^2} \tag{2.19}$$

2.2.1.3 Merging and Splitting Poisson Arrival Processes

Let us now consider two customers sets, called, respectively, A and B. Let us assume the customers in the two sets are independent, therefore the behavior of customers in set A is not influenced by customers in set B and vice versa. Customers in A

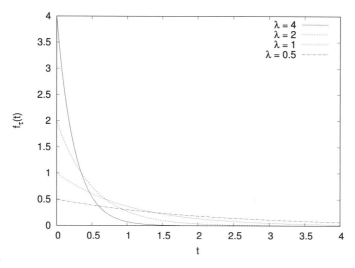

Fig. 2.5 The negative exponential probability density function plotted for some values of λ

send requests as a Poisson process with rate λ_A; customers in B send requests as a Poisson process with rate λ_B. Now assume that the requests from costumer set A and customer set B are all sent to the same queuing system in a way that makes them indistinguishable (i.e., the system cannot tell whether a request comes from a customer in set A or B). As far as the queuing system is concerned, the arrival process is a single one resulting from the *merging* of the two arrival processes from the two customers sets, i.e., the queuing system sees just one single arrival process summing up the arrivals from the two customers sets, as shown in the graphical example of Fig. 2.6.

Now the question is: can we tell something about the statistics of the new process resulting by merging two Poisson processes? Indeed we can say that the average arrival rate will be the sum of the two individual averages, meaning that if λ is the average arrival rate of the merged process, then:

$$\lambda = \lambda_A + \lambda_B \tag{2.20}$$

This result is true in general and is also intuitively easy to explain since the number of customers just sum up. Therefore, if on average 3 customers arrive from set A and 7 from set B in a time unit, the average total number of customers arriving in the time unit will be 10, as in Fig. 2.6.

It is by far less intuitive to make a guess about the general statistics of the process, such as the distribution function. To get an insight on this, let us proceed as we did above to derive equations (2.8) and (2.9). Let us consider a sample interval T and divide it in N sub-intervals of equal size ΔT. We can write:

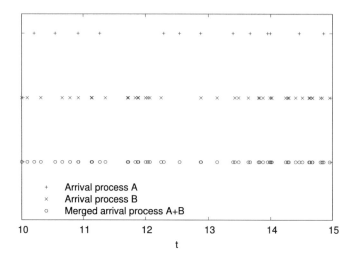

Fig. 2.6 Merging Poisson processes: the merging of two Poisson processes with $\lambda_A = 3$ and $\lambda_B = 7$ gives birth to a Poisson process with average arrival rate $\lambda = 10$

$$\begin{aligned}
\Pr\{1 \text{ arrival from customer set A}\} &= \lambda_A \Delta T \\
\Pr\{1 \text{ arrival from customer set B}\} &= \lambda_B \Delta T
\end{aligned} \tag{2.21}$$

Since processes A and B are independent, it follows that:

$$\begin{aligned}
\Pr\{\text{no arrivals in } \Delta T\} &= \Pr\{\text{no arrivals from A}\} \Pr\{\text{no arrivals from B}\} \\
&= (1 - \lambda_A \Delta T)(1 - \lambda_B \Delta T) \\
&= 1 - (\lambda_A + \lambda_B)\Delta T + \lambda_A \lambda_B \Delta T^2 \\
&= 1 - (\lambda_A + \lambda_B)\Delta T + o(\Delta T)
\end{aligned} \tag{2.22}$$

$$\begin{aligned}
\Pr\{1 \text{ arrival in } \Delta T\} &= \Pr\{\text{no arrivals from A}\} \Pr\{1 \text{ arrival from B}\} \\
&\quad + \Pr\{1 \text{ arrival from A}\} \Pr\{\text{no arrivals from B}\} \\
&= (1 - \lambda_A \Delta T)\lambda_B \Delta T + \lambda_A \Delta T(1 - \lambda_B \Delta T) \\
&= (\lambda_A + \lambda_B)\Delta T + o(\Delta T)
\end{aligned} \tag{2.23}$$

$$\begin{aligned}
\Pr\{2 \text{ arrivals in } \Delta T\} &= \Pr\{1 \text{ arrival from A}\} \Pr\{1 \text{ arrival from B}\} \\
&= \lambda_A \Delta T \lambda_B \Delta T = o(\Delta T)
\end{aligned} \tag{2.24}$$

Equation (2.24) says that the probability of multiple arrivals in ΔT is negligible when compared with the probability of zero arrivals or one arrival. Therefore, we may focus on these two meaningful cases only, which happen with probabilities:

$$P_0 = 1 - (\lambda_A + \lambda_B)\Delta T$$
$$P_1 = (\lambda_A + \lambda_B)\Delta T$$

(2.25)

Starting from (2.25) and applying the limit for $N \to \infty$, similarly to what done in (2.11), we get that the probability of k arrivals in T is still given by the Poisson formula with rate $\lambda = \lambda_A + \lambda_B$. This can be easily generalized to any number of merged Poisson processes, leading to the following general result:

The process resulting after merging multiple Poisson processes is still a Poisson process with average arrival rate equal to the sum of the arrival rates of the merged processes, such that

$$\lambda = \sum_{\forall i} \lambda_i$$

(2.26)

As two Poisson processes can be merged into one Poisson process, it is also possible to split a single Poisson process into two different processes. Let us assume that we want to randomly split an original single arrival process into process A and process B. When a new arrival occurs, it is randomly assigned to arrival process A with probability P_A and to process B with probability P_B. Assuming that $P_A + P_B = 1$, the total number of arrivals will still be the same as in the original process.

Now let us extend this idea to splitting an original Poisson process into N children processes with probabilities $P_1, P_2, \ldots, P_i, \ldots P_N$. It may be shown that the N children processes are still Poisson processes.

The N processes resulting from the random split of a Poisson process according to a set of probabilities $P_1, P_2, \ldots P_i, \ldots P_N$ are still Poisson processes with arrival rates

$$\lambda_i = P_i \lambda \quad \forall i \in [1; N]$$

(2.27)

2.3 Modeling Service Time

Once the arrival process to the queuing system is modeled, the same has to be done for the service time, which is also a random quantity. In this case the most straightforward way is to look at the probability distribution function (or the probability density function) of the time required by a generic server to serve one of the customers that entered the system.

Given a queuing system, let us assume that the *service time* is a random variable ϑ with a finite mean $\bar{\vartheta}$, which will be called *average service time* in the following. The (average) service time is measured in time units, such as seconds or minutes. The *average service rate* of a server will then be defined as the average rate at which services are completed by the server and is given by:

$$\mu = \frac{1}{\bar{\vartheta}} \qquad (2.28)$$

In the following we will consider a few service time distributions that can be useful to represents services typically offered by teletraffic systems, namely exponential, deterministic, uniform, Erlang, and Pareto distributions.

The exponential service time distribution found widespread application in the engineering of telephone networks. The assumption of exponential service time is of course an approximation, as it implies that very short service times are most likely, while some large service times can reach almost infinity. Daily life experience tells us that this is not the case: most of the calls have a duration of a few minutes, very rarely they last a few seconds and almost never for a very long time. Nonetheless, this modeling approach was very effective since the results obtained found good correspondence with real-life system behaviors. Therefore the assumption of exponential service times dominates most of the teletraffic work devoted to telephone networks.

When coming to packet-switched networks the situations changes. Packets usually have pre-determined lower and upper bounds in length or must have a fixed length. Therefore, the exponential model may clearly lack accuracy when dealing with packet-switched networks. For this reason in the part of this book devoted to packet-switched networks we will deal not only with the exponential distribution, which will give us the opportunity to use the modeling tools already developed for the telephone networks, but we will also consider other distributions that may better model typical packet length distributions.

2.3.1 *Exponential Service Time*

A possible assumption, in analogy with the Poisson inter-arrival time, is that the service time is exponentially distributed. Therefore, we can use Eqs. (2.16) and (2.17) and define the service time probability distribution and density functions as:

$$F_\vartheta(t) = \begin{cases} 0 & t < 0 \\ 1 - e^{-\mu t} & t \geq 0 \end{cases}$$

$$f_\vartheta(t) = \begin{cases} 0 & t < 0 \\ \mu e^{-\mu t} & t \geq 0 \end{cases} \tag{2.29}$$

The average and variance of the exponential service time are, respectively:

$$\bar{\vartheta} = \frac{1}{\mu} \tag{2.30}$$

$$\sigma_\vartheta^2 = \frac{1}{\mu^2} \tag{2.31}$$

2.3.2 Deterministic Service Time

Another possibility that finds application and will be used in the following is to assume that the service time has a fixed duration ϑ_0, which is the same for all customers. In this case the probability distribution function is:

$$F_\vartheta(t) = \begin{cases} 0 & t < \vartheta_0 \\ 1 & t \geq \vartheta_0 \end{cases} \tag{2.32}$$

The average and variance of the deterministic service time are, respectively:

$$\bar{\vartheta} = \vartheta_0 \tag{2.33}$$

$$\sigma_\vartheta^2 = 0 \tag{2.34}$$

2.3.3 Uniform Service Time

If the service time takes random values between a minimum value ϑ_m and a maximum value ϑ_M, it has a uniform distribution ($\mathscr{U}$ in the Kendall notation). In this case the probability density function is expressed as:

$$f_\vartheta(t) = \begin{cases} 0 & t < \vartheta_m \text{ and } t > \vartheta_M \\ \frac{1}{\vartheta_M - \vartheta_m} & \vartheta_m \leq t \leq \vartheta_M \end{cases} \tag{2.35}$$

The probability distribution function can be obtained as:

$$F_\vartheta(t) = \int_{-\infty}^{t} f_\vartheta(u)\,du = \begin{cases} 0 & t < \vartheta_m \\ \frac{t-\vartheta_m}{\vartheta_M - \vartheta_m} & \vartheta_m \le t \le \vartheta_M \\ 1 & t > \vartheta_M \end{cases} \tag{2.36}$$

The average and variance of the uniform service time are, respectively:

$$\bar{\vartheta} = \frac{\vartheta_M + \vartheta_m}{2} \tag{2.37}$$

$$\sigma_\vartheta^2 = \frac{(\vartheta_M - \vartheta_m)^2}{12} \tag{2.38}$$

2.3.4   Erlang Service Time

Another description of the service time that may have some interesting practical applications is the one that adopts the Erlang distribution. Let us assume that the total service time ϑ has a value resulting from the sum of a series of r smaller time periods $\vartheta_1, \vartheta_2, \ldots, \vartheta_r$ that are all independent and identically distributed according to an exponential distribution with the same average value $\bar{\vartheta}_0$. Such a composition of the service time is useful to represent the case of a service consisting of multiple stages, such as the one provided by a pipeline processing system. We can write:

$$\vartheta = \sum_{i=1}^{r} \vartheta_i \quad \text{and} \quad \bar{\vartheta} = \frac{1}{\mu} = r\bar{\vartheta}_0 \tag{2.39}$$

In this case the probability density function of the service time is given by the convolution of r exponential probability density functions in the form $f_{\vartheta_i}(t) = r\mu e^{-r\mu t}$, resulting in the following expression:

$$f_{r,\vartheta}(t) = \begin{cases} 0 & t < 0 \\ \frac{(r\mu t)^{r-1}}{(r-1)!} r\mu e^{-r\mu t} & t \ge 0 \end{cases} \tag{2.40}$$

The probability distribution function is:

$$F_{r,\vartheta}(t) = \begin{cases} 0 & t < 0 \\ 1 - e^{-r\mu t} \sum_{j=0}^{r-1} \frac{(r\mu t)^j}{j!} & t \ge 0 \end{cases} \tag{2.41}$$

This is called an Erlang distribution of grade r. The variance of the Erlang-distributed service time is:

$$\sigma_\vartheta^2 = r\bar{\vartheta}_0^2 \tag{2.42}$$

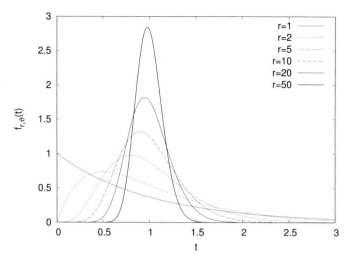

Fig. 2.7 The Erlang probability density function of grade r for some values of r and $\bar{\vartheta}_0 = 1/r$

Figure 2.7 shows the Erlang probability density function of grade r for some values of r and $\bar{\vartheta}_0 = 1/r$, such that the average service time is always $\bar{\vartheta} = 1$. For $r = 1$ the curve becomes an exponential probability density function with $\mu = 1$, whereas when r increases the density function tends to a very sharp curve centered on the average value. It can be proved that for $r \rightarrow \infty$ the Erlang distribution becomes a deterministic distribution.

2.3.5 Pareto Service Time

The Poisson process has been widely used in basic teletraffic engineering with good results. Nonetheless, in recent years other models have been considered by the scientific community. In particular, it was discovered that the data traffic observed in a LAN is not well described by a Poisson process, mainly because it exhibits what is called a "long range dependent" behavior.

A detailed analysis of such a phenomenon is well beyond the scopes of this book, but in brief we can say that this property can be captured by assuming the service time distributed according to the Pareto distribution, which in its most elementary form has the following probability distribution and density functions:

$$F_\vartheta(t) = \begin{cases} 0 & t < t_0 \\ 1 - \left(\frac{t_0}{t}\right)^\alpha & t \geq t_0 \end{cases} \tag{2.43}$$

$$f_\vartheta(t) = \begin{cases} 0 & t < t_0 \\ \dfrac{\alpha t_0^\alpha}{t^{\alpha+1}} & t \geq t_0 \end{cases} \tag{2.44}$$

where $t_0 > 0$ and $\alpha > 0$ are called scale parameter and shape parameter, respectively.

It is very interesting to compute average and variance of the Pareto-distributed service time, given by:

$$\bar\vartheta = \begin{cases} \infty & \alpha \leq 1 \\ \dfrac{\alpha t_0}{\alpha-1} & \alpha > 1 \end{cases} \tag{2.45}$$

$$\sigma_\vartheta^2 = \begin{cases} \infty & \alpha \leq 2 \\ \left(\dfrac{t_0}{\alpha-1}\right)^2 \dfrac{\alpha}{\alpha-2} & \alpha > 2 \end{cases} \tag{2.46}$$

In particular, we can see that for $1 < \alpha \leq 2$ it is possible to have a finite average value of the service time with an infinite variance, meaning that the service time is extremely variable around its average. Under this condition the Pareto probability density function is said to be "heavy-tailed." The reason is that it goes to 0 when $t \rightarrow \infty$ with a polynomial trend, and the tail of the distribution is less negligible with respect to the values concentrated around the average, if we compare it, for instance, with an exponential distribution. This is clearly visible from Fig. 2.8, where the Pareto probability density function is plotted in logarithmic scale for some values of α and t_0 such that $\bar\vartheta = 1$, and compared with the exponential probability density function with $\mu = 1$.

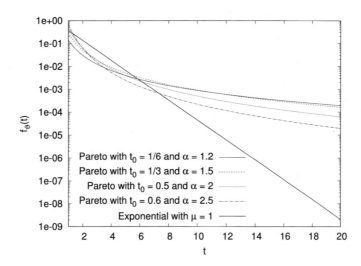

Fig. 2.8 The Pareto probability density function for some values of α and t_0 such that $\bar\vartheta = 1$. Comparison with the negative exponential probability density function with $\mu = 1$

2.4 Link the Time of Arrivals with the Service Time

In this section we briefly address the relationship between arrival time and service time in the traffic model we will consider in this book. The Poisson process describes the behavior of the arrival instances, i.e., how they are randomly distributed in time, while the service time model describes how long a given customer will keep the server busy. This two models are at this stage fully un-correlated, therefore there are no links between them.

In practice there are no assumptions on *how* the traffic is physically generated, and it is assumed that any possible mapping of service time with arrival instants is possible. To help understanding this statement let us consider Fig. 2.9. It plots Poisson arrivals generated as in Fig. 2.6, but adding the related service time as a bar. The service times are drawn randomly according to an exponential distribution without any correlation with the specific arrival time of the customer and of the previous and following customers. In real systems the service time is known only at the end of the service when the customer leaves the system. However, here it is plotted as if it was known at the time of arrival. We can easily see that service times are mixed up randomly and there is no specific correlation with the time of arrival.

This is an important concept that we will recall later in Chap. 5 where we will better understand its implications.

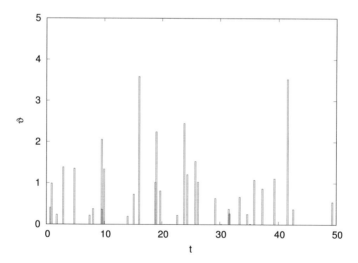

Fig. 2.9 Graphical description of a traffic flow generated as a Poisson arrival process and exhibiting exponential service time. The length of the bars gives the specific service time of the customer arriving at that particular time instant

2.5   Residual Service Time

Given that a customer exhibits a random service time according to some kind of distribution, it will start being served at the generic time instant t_0 and will stop being served at time $t_0 + \vartheta$. Now let us consider a server serving customers within a system with some specific behavior and imagine that, at some random time instant, an external observer looks at the system and at that particular server. If the server is busy, the observer will observe a service period that lasts a specific value ϑ_s. We use the subscript s to highlight the fact that this is not a generic service time, but a service time that has been *sampled* by the observer. Let us call t_1 the instant when the observer starts observing the sampled service period, obviously with $t_0 \leq t_1 \leq t_0 + \vartheta_s$. That particular service period already lasted $t_1 - t_0$ and will last some additional time up to $t_0 + \vartheta_s$, which is a random quantity from the observer's perspective.

Let us call this remaining service time ζ as the *residual service time*, as illustrated in Fig. 2.10, and let us ask ourselves whether it is possible to tell something about the statistics of ϑ_s and ζ, given that we have some knowledge of the statistics of ϑ. It is important to clarify that also the sampled service time ϑ_s is a random variable and that in general it is different from ϑ, as we will show in a moment.

In general terms, it is easy to understand that something can be done to answer that question. If, for instance, we assume a deterministic service time such that $\vartheta = \vartheta_0$, we know for sure that the observer will always sample a service period of length $\vartheta_s = \vartheta_0$ and that the residual service time will have the value $\zeta = \vartheta_0 - t_1 + t_0$. Now the question is whether this simple idea can be generalized to any service time distribution.

The most important observation to this end regards ϑ_s. It is important to notice that the external observer chooses a random instant in time to start observing the sampled service period. Therefore, it is more likely that the observer chooses an instant belonging to a long service time rather than to a short one. For instance, let us imagine that the service time can have either duration $\vartheta_1 = 1$ min or $\vartheta_2 = 2$ min. The random observer has twice as many chances to sample a service of length ϑ_2

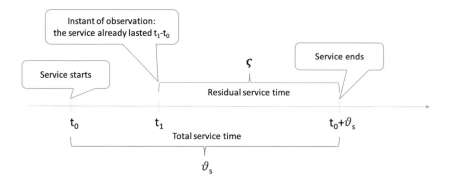

Fig. 2.10 Graphical representation of the residual service time

than a service of length ϑ_1. Following and generalizing this observation, we can say that the probability to sample a specific value ϑ_s of the service time is proportional to ϑ_s itself. Let us use the generic variables t and s to indicate possible values of ζ and ϑ_s, respectively. Therefore, we can write:

$$f_{\vartheta_s}(s) = Csf_\vartheta(s) \quad s > 0 \tag{2.47}$$

with C being a constant that must be calculated to make the integral of the probability density function equal to 1, that is:

$$\int_0^\infty f_{\vartheta_s}(s)ds = 1 \tag{2.48}$$

The result is $C = 1/\bar{\vartheta}$ and therefore:

$$f_{\vartheta_s}(s) = \frac{sf_\vartheta(s)}{\bar{\vartheta}} \quad s > 0 \tag{2.49}$$

Now we can compute the probability density function of ζ subject to the fact that the observer sampled a given $\vartheta_s = s$. It follows that the value of the residual service time t may be any between t_0 and $t_0 + s$ with a uniform random distribution, so that:

$$f_{\zeta|\vartheta_s}(t \mid s) = \begin{cases} \frac{1}{s} & 0 < t \le s \\ 0 & \text{elsewhere} \end{cases} \tag{2.50}$$

Therefore:

$$f_{\zeta,\vartheta_s}(t,s) = f_{\vartheta_s}(s)f_{\zeta|\vartheta_s}(t \mid s) = \frac{sf_\vartheta(s)}{s \cdot \bar{\vartheta}} = \frac{f_\vartheta(s)}{\bar{\vartheta}} \quad t > 0 \text{ and } t \le s \tag{2.51}$$

Obviously $\zeta \le \vartheta_s$ i.e., $t \le s$ since the residual service time cannot be longer than the original one and we can then calculate the probability density function of ζ by applying the theorem of the total probability:

$$f_\zeta(t) = \int_{-\infty}^\infty f_{\zeta,\vartheta_s}(t,s)ds = \int_t^\infty \frac{f_\vartheta(s)}{\bar{\vartheta}}ds = \begin{cases} \frac{1-F_\vartheta(t)}{\bar{\vartheta}} & t \ge 0 \\ 0 & t < 0 \end{cases} \tag{2.52}$$

Now that we know the probability density function of ζ, it is possible to calculate its average:

$$\bar{\zeta} = \int_0^\infty tf_\zeta(t)dt = \int_0^\infty t\frac{1-F_\vartheta(t)}{\bar{\vartheta}}dt = \frac{1}{\bar{\vartheta}}\int_0^\infty t(1-F_\vartheta(t))dt$$

$$= \frac{1}{\bar{\vartheta}}\left[\left|\frac{t^2}{2}(1-F_\vartheta(t))\right|_0^\infty + \int_0^\infty \frac{t^2}{2}f_\vartheta(t)dt\right] = \frac{E[\vartheta^2]}{2\bar{\vartheta}} = \frac{\sigma_\vartheta^2 + \bar{\vartheta}^2}{2\bar{\vartheta}} \tag{2.53}$$

assuming that $\lim_{t \to \infty} t^2 (1 - F_\vartheta(t)) = 0$. More generally, the k-th moment of ζ is a function of $\bar{\vartheta}$ and of the $k + 1$-th moment of ϑ:

$$E[\zeta^k] = \int_0^\infty t^k f_\zeta(t) dt = \int_0^\infty t^k \frac{1 - F_\vartheta(t)}{\bar{\vartheta}} dt = \frac{1}{\bar{\vartheta}} \int_0^\infty t^k (1 - F_\vartheta(t)) dt$$

$$= \frac{1}{\bar{\vartheta}} \left[\left| \frac{t^{k+1}}{k+1} (1 - F_\vartheta(t)) \right|_0^\infty + \int_0^\infty \frac{t^{k+1}}{k+1} f_\vartheta(t) dt \right] = \frac{E[\vartheta^{k+1}]}{(k+1)\bar{\vartheta}}$$

(2.54)

assuming that $\lim_{t \to \infty} t^{k+1} (1 - F_\vartheta(t)) = 0$.

2.5.1   The Residual Exponential Service Time and Its Memoryless Property

For the exponential probability density function, starting from (2.52), we obtain that:

$$f_\zeta(t) = \frac{1 - F_\vartheta(t)}{\bar{\vartheta}} = \frac{1 - 1 + e^{-\mu t}}{\bar{\vartheta}} = \mu e^{-\mu t} \quad t \geq 0 \qquad (2.55)$$

It turns out that for exponential services the probability density function of the residual service time is exactly equal to the probability density function of the service itself. This is equivalent to say that, even though part of the service time has already expired, this fact does not influence in any way the statistics of what remains of the service time. This is a very important property called the *memoryless* property of the exponential distribution, which is the only random distribution that exhibits this property. The memoryless nature of the exponential service time has a very important impact on the modeling tools that will be presented in the next chapters.

The same property can be also demonstrated directly by recalling that the probability of an event A under the condition of another event B can be calculated from the joint probability of both events A and B as follows:

$$\Pr\{A|B\} = \frac{\Pr\{A, B\}}{\Pr\{B\}} \qquad (2.56)$$

Now let us say that event A is the residual service time ζ to be less than a given value t, and event B is the total service time ϑ to be greater than the part of the service time v that has already expired. Then:

$$\Pr\{A|B\} = \Pr\{\zeta \le t \mid \vartheta \ge v\} = F_\zeta(t) \tag{2.57}$$

since this formula basically represents the definition of the probability distribution function of the residual service time $\zeta = \vartheta - v \le t$.

Then we can observe that:

$$\Pr\{A, B\} = \Pr\{v \le \vartheta \le t + v\} = 1 - e^{-\mu(t+v)} - 1 + e^{-\mu v} = e^{-\mu v} - e^{-\mu(t+v)} \tag{2.58}$$

and that:

$$\Pr\{B\} = \Pr\{\vartheta \ge v\} = 1 - 1 + e^{-\mu v} \tag{2.59}$$

We can then write:

$$F_\zeta(t) = \frac{e^{-\mu v} - e^{-\mu(t+v)}}{e^{-\mu v}} = 1 - e^{-\mu t} \tag{2.60}$$

Therefore, the residual service time has the same probability distribution as the service time.

2.5.2   The Residual Deterministic Service Time

Deterministic service time means that all service times have the same value $\vartheta = \vartheta_0$, equal to the average value. In this case, when an observer arrives at random during a service time he/she will choose any instant t_1 between t_0 and $t_0 + \vartheta_0$ with equal probability. Therefore, the service time that has already expired $t_1 - t_0$ is uniformly distributed between 0 and ϑ_0. Being ϑ_0 constant, the residual service time $\zeta = \vartheta_0 - (t_1 - t_0)$ is also uniformly distributed between 0 and ϑ_0, with probability density function:

$$f_\zeta(t) = \begin{cases} \frac{1}{\vartheta_0} & \text{for } 0 \le t \le \vartheta_0 \\ 0 & \text{otherwise} \end{cases} \tag{2.61}$$

It follows that:

$$\bar\zeta = \int_0^{\vartheta_0} t \frac{1}{\vartheta_0} dt = \frac{\vartheta_0}{2} \tag{2.62}$$

which we can also derive considering that $E[\vartheta^2] = \vartheta_0^2$ and therefore:

$$\bar\zeta = \frac{E[\vartheta^2]}{2\bar\vartheta} = \frac{\vartheta_0^2}{2\vartheta_0} = \frac{\vartheta_0}{2} \tag{2.63}$$

2.5.3 The Residual Uniform Service Time

Let us assume that the service time has a uniform distribution between ϑ_m and ϑ_M. From Eqs. (2.36), (2.37), and (2.52) it follows that:

$$f_\zeta(t) = \begin{cases} \frac{2}{\vartheta_M + \vartheta_m} & \text{for } 0 \le t \le \vartheta_m \\ 2\frac{\vartheta_M - t}{\vartheta_M^2 - \vartheta_m^2} & \text{for } \vartheta_m \le t \le \vartheta_M \\ 0 & \text{otherwise} \end{cases} \tag{2.64}$$

and therefore we can calculate:

$$\bar{\zeta} = \int_0^\infty t f_\zeta(t)\, dt = \int_0^{\vartheta_m} \frac{2t}{\vartheta_M + \vartheta_m}\, dt + \int_{\vartheta_m}^{\vartheta_M} 2t\frac{\vartheta_M - t}{\vartheta_M^2 - \vartheta_m^2}\, dt \tag{2.65}$$

$$= \frac{\vartheta_M}{3} + \frac{\vartheta_m^2}{3(\vartheta_M + \vartheta_m)}$$

which we can also derive from (2.37) and (2.38) as follows:

$$\bar{\zeta} = \frac{\sigma_\vartheta^2 + \bar{\vartheta}^2}{2\bar{\vartheta}} = \frac{1}{\vartheta_M + \vartheta_m}\left(\frac{(\vartheta_M - \vartheta_m)^2}{12} + \frac{(\vartheta_M + \vartheta_m)^2}{4}\right) \tag{2.66}$$

$$= \frac{\vartheta_M}{3} + \frac{\vartheta_m^2}{3(\vartheta_M + \vartheta_m)}$$

2.6 Examples and Case Studies

2.6.1 Time Related Tariffs

Let us assume that a given telecommunication service is charged on the basis of the amount of time the customer uses it. The service may be, for instance, a mobile phone call or a dial-up connection to the Internet. Let us assume that the duration of the service usage ϑ is random with average value $\bar{\vartheta} = 1/\mu$ and exponential probability density function $f_\vartheta(t) = \mu e^{-\mu t}$.

Very often the charging policy that determines the cost of the service is a consequence of a *counting* process that gives, for every service, a number n_t of *tokens* as follows:

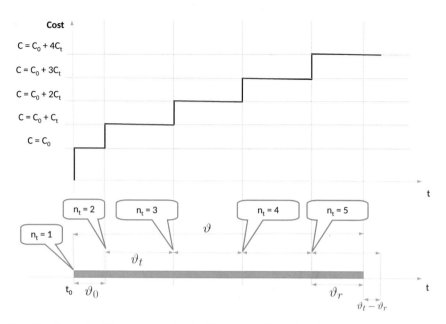

Fig. 2.11 Example of the costs associated with a given call as a function of its duration ϑ. In this example $\vartheta_t = 20\,\text{s}$, $\vartheta_0 = 10\,\text{s}$, $C_t = 0.10$ and $C_0 = 0.20$, therefore $\alpha = 2$

1. $n_t = 0$ just before the service starts
2. $n_t = n_t + 1$ as the service starts (this is also called the "calling charge")
3. $n_t = n_t + 1$ when a first charging slot of size ϑ_0 expires;
4. $n_t = n_t + 1$ every time a new charging slot of value ϑ_t expires.

The cost C of a call is determined as a function of n_t. A strategy that is often adopted, for a reason that will be clarified later, is to charge a specific cost C_0 to the first token and a fixed cost C_t to all the subsequent tokens. The whole process is summarized in Fig. 2.11.

We are interested in studying the average call pricing $\bar{C}$ as a function of ϑ_0, ϑ_t, C_0, and C_t:

$$\bar{C} = C_0 + C_t(\bar{n}_t - 1) \tag{2.67}$$

where $\bar{n}_t$ is the average number of tokens expressed as a function of ϑ_0 and ϑ_t. Given the exponential distribution of ϑ, it follows that n_t is a discrete random variable that can take any positive integer value. Let us call $P(n_t)$ the probability of a given value of n_t:

$$
\begin{aligned}
P(0) &= 0 \\
P(1) &= \Pr\{\vartheta < \vartheta_0\} = 1 - e^{-\mu\vartheta_0} \\
P(k > 1) &= \Pr\{\vartheta_0 + (k-2)\vartheta_t \le \vartheta < \vartheta_0 + (k-1)\vartheta_t\} \\
&= e^{-\mu[\vartheta_0 + (k-2)\vartheta_t]} - e^{-\mu[\vartheta_0 + (k-1)\vartheta_t]}
\end{aligned}
\tag{2.68}
$$

Therefore

$$
\begin{aligned}
\bar{n}_t &= \sum_{k=0}^{\infty} n_t P(n_t) = 1 - e^{-\mu\vartheta_0} + \sum_{k=2}^{\infty} k\left(e^{-\mu[\vartheta_0 + (k-2)\vartheta_t]} - e^{-\mu[\vartheta_0 + (k-1)\vartheta_t]}\right) \\
&= 1 - e^{-\mu\vartheta_0} + e^{-\mu\vartheta_0} \sum_{k=2}^{\infty} k\left(e^{-\mu[(k-2)\vartheta_t]} - e^{-\mu[(k-1)\vartheta_t]}\right) \\
&= 1 - e^{-\mu\vartheta_0} + e^{-\mu\vartheta_0}\left[2(1 - e^{-\mu\vartheta_t}) + 3(e^{-\mu\vartheta_t} - e^{-\mu2\vartheta_t}) + 4\left(e^{-\mu2\vartheta_t} - e^{-\mu3\vartheta_t}\right) + \ldots\right] \\
&= 1 - e^{-\mu\vartheta_0} + e^{-\mu\vartheta_0}\left(2 + e^{-\mu\vartheta_t} + e^{-2\mu\vartheta_t} + e^{-3\mu\vartheta_t} + \ldots\right) \\
&= 1 - e^{-\mu\vartheta_0} + e^{-\mu\vartheta_0}\left(1 + \sum_{i=0}^{\infty} e^{-i\mu\vartheta_t}\right) = 1 - e^{-\mu\vartheta_0} + e^{-\mu\vartheta_0}\left(1 + \frac{1}{1 - e^{-\mu\vartheta_t}}\right) \\
&= 1 + \frac{e^{-\mu\vartheta_0}}{1 - e^{-\mu\vartheta_t}}
\end{aligned}
\tag{2.69}
$$

Assuming that ϑ_t is fixed, let us then compare the following two cases:

1. ϑ_0 is a deterministic value between 0 and ϑ_t;
2. ϑ_0 is a random variable taking values between 0 and ϑ_t with a uniform distribution.[4]

In the former case, formula (2.69) returns a constant value for $\bar{n}_t$. If we choose $\vartheta_0 = \vartheta_t$, we get:

$$
\bar{n}_t = 1 + \frac{e^{-\mu\vartheta_t}}{1 - e^{-\mu\vartheta_t}} = \frac{1}{1 - e^{-\mu\vartheta_t}}
\tag{2.70}
$$

In case ϑ_0 is random, formula (2.69) returns $\bar{n}_t(\vartheta_0)$ as a function of the random variable ϑ_0, which has probability density function:

---

[4]This case resulted from a method that was used in the old analog telephone network, where it was not possible to provide a specific counter per call and a single counter was available for a given central office, triggering counting steps periodically at ϑ_t pace. In that situation the number of tokens per call was given by the number of counting steps of the general counter that happened during the call, such that the first counting step occurred randomly after the start of the call, that is a purely random quantity between 0 and ϑ_t. For instance, in Italy this token generation algorithm was required by the law and is described in art (10 of the Decree of the Minister of Telecommunications of 28/2/97).

$$f_{\vartheta_0}(t) = \begin{cases} \frac{1}{\vartheta_t} & 0 \le t \le \vartheta_t \\ 0 & t < 0 \text{ or } t > \vartheta_t \end{cases} \tag{2.71}$$

By averaging the function of the random variable we can obtain $\bar{n}_t$ independently of ϑ_0 as follows:

$$\bar{n}_t = \int_0^{\vartheta_t} \frac{1}{\vartheta_t} \left[1 + \frac{e^{-\mu t}}{1 - e^{-\mu \vartheta_t}} \right] dt$$

$$= 1 + \frac{1}{\vartheta_t (1 - e^{-\mu \vartheta_t})} \int_0^{\vartheta_t} e^{-\mu t} dt = 1 + \frac{1}{\mu \vartheta_t} = 1 + \frac{\bar{\vartheta}}{\vartheta_t} \tag{2.72}$$

2.6.1.1 Calculating the Average Cost of the Calls

Let us assume that the time unit used to calculate the call duration is the second. In general terms today computers may measure the time with greater accuracy, but indeed for human related activities the second is a quite accurate measurement unit. Under this assumption, let us call C_s the cost of the call per second. Then we expect that a call of average duration $\bar{\vartheta}$ seconds will cost exactly:

$$C = C_0 + \bar{\vartheta} C_s \tag{2.73}$$

Now the question is how to calculate the cost of the call when its duration is not expressed in seconds, but tokens are be used for charging, as outlined above. The average cost per second should be the same in principle, and the most intuitive way to compute it is to calculate the cost per token according to the duration of the interval between tokens:[5]

$$C_t = \vartheta_t C_s \tag{2.74}$$

Now let us discuss the process of calculating the cost of the call using the tokens. Since the cost of the call is given by Eq. (2.67), using Eqs. (2.69) and (2.72) we obtain:

$$\bar{C} = \begin{cases} C_0 + C_t \frac{e^{-\mu \vartheta_0}}{1 - e^{-\mu \vartheta_t}} & \vartheta_0 \text{ deterministic} \\ C_0 + C_t \frac{\bar{\vartheta}}{\vartheta_t} = C_0 + C_s \bar{\vartheta} & \vartheta_0 \text{ random} \end{cases} \tag{2.75}$$

Figure 2.12 shows the cost $\bar{C}$ as a function of ϑ_t for different deterministic values of ϑ_0. The figure shows that the cost of the call is lower for larger values of the first

---

[5]This is what is usually done in typical tariffs for mobile phones or fixed phones in the countries where this is applicable.

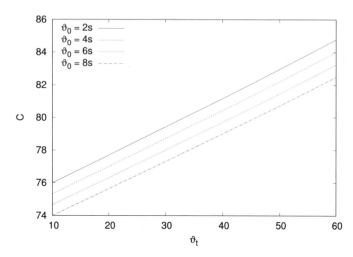

Fig. 2.12 Average call pricing $\bar{C}$ as a function of ϑ_t, for $\bar{\vartheta} = 180$ s, $C_0 = 15$ cents, $C_s = 1/3$ cent/s, for different deterministic values of ϑ_0

charging slot ϑ_0, which is intuitive, but also that $\bar{C}$ increases with the size ϑ_t of the following charging slots, which is quite unexpected. This behavior can be explained by considering that the cost calculation based on tokens "rounds up" the duration of a call to a multiple of ϑ_t. If we observe Fig. 2.11 we can see that, since the call may have any duration ϑ, the residual call time after the last generated token is $\vartheta_r < \vartheta_t$ and should contribute to the cost for $\vartheta_r C_s$. However, since a new token is generated only after an interval ϑ_t expires, the residual call time ϑ_r is not actually charged, thus giving a potential advantage to the customer. Such an advantage is therefore compensated by setting the first counting interval ϑ_0 to a value that is smaller than ϑ_t. If $\vartheta_0 \ll \vartheta_t$, in practice all counting intervals are charged "in advance" and the residual call time is charged as a full token, i.e., with a cost $C_t = \vartheta_t C_s > \vartheta_r C_s$. This effect becomes more relevant when ϑ_t increases, as shown in Fig. 2.12, since the charging scheme rounds up the duration of the call charged to the user to a larger value.

Figure 2.13 explains the role of ϑ_0 in determining the call price, as outlined above. The figure shows that the operator has an economic advantage on the user when ϑ_0 is small, whereas the user has an advantage when ϑ_0 is large. It is quite interesting to note that choosing ϑ_0 randomly results in fair cost charging, because when ϑ_0 is random it perfectly balances, on average, the residual call time ϑ_r not charged at the end of the call.

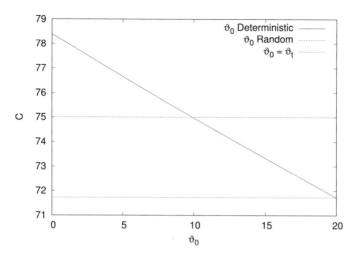

Fig. 2.13 Average call pricing $\bar{C}$ as a function of ϑ_0, for $\bar{\vartheta} = 180$ s, $C_0 = 15$ cents, $C_s = 1/3$ cent/s, and $\vartheta_t = 20$ s. Comparison among different choices for ϑ_0

2.6.2 Deriving the Poisson Formula from Exponential Inter-arrivals

In Sect. 2.2.1.2, starting from the Poisson formula we proved that the inter-arrival times of a Poisson process are exponentially distributed. Can we say the opposite, i.e., that a process with independent and identically distributed exponential inter-arrival times with average $\bar{\tau} = \frac{1}{\lambda}$ is indeed a Poisson arrival process? In this section we show the reverse proof of the link existing between the negative exponential distribution and the Poisson formula of the probability of k arrivals in an interval of length T.

Let us start the observation of the arrival process at time t_0 and let us call

- t_n the time of the n-th arrival;
- k the total number of arrivals at time t, where $t = t_0 + T$;
- $P(k, T)$ the probability of k arrivals in T.

Considering that there are k arrivals in T if and only if $t_k \leq t$ and $t_{k+1} > t$, we can write:

$$P(k, T) = \Pr(t_{k+1} > t) - \Pr(t_k > t) \tag{2.76}$$

If we find $\Pr(t_k > t)$, we can derive $P(k, T)$. We can calculate the probability distribution function of $t_k - t_0$ by observing that it is the sum of k inter-arrival times that are independent and identically distributed according to an exponential distribution with parameter λ. This is equivalent to say that $t_k - t_0$ follows an Erlang distribution of grade k with average k/λ. Therefore, from Eq. (2.41) it follows:

$$\Pr(t_k \le t) = F_{t_k - t_0}(T) = 1 - e^{-\lambda T} \sum_{j=0}^{k-1} \frac{(\lambda T)^j}{j!} \tag{2.77}$$

In conclusion, from Eq. (2.76) we can derive the Poisson formula as:

$$P(k, T) = \left(1 - 1 + e^{-\lambda T} \sum_{j=0}^{k} \frac{(\lambda T)^j}{j!}\right) - \left(1 - 1 + e^{-\lambda T} \sum_{j=0}^{k-1} \frac{(\lambda T)^j}{j!}\right) = e^{-\lambda T} \frac{(\lambda T)^k}{k!}$$
$$\tag{2.78}$$

2.6.3　Time to Complete Multiple Services

Let us consider a system with m servers, assuming that the service times required by the customers are independent and identically distributed. We are interested in comparing two cases:

1. the service times are deterministic with value ϑ_0;
2. the service times are exponentially distributed with average $\bar{\vartheta} = \vartheta_0$.

At time t_0 the m servers start serving a customer each, for a total of m customers being served, and no more customers join the system. We are interested in knowing when the system will be completely free. In particular we want to calculate:

- the average time $\bar{\vartheta}_f$ when the first customer leaves the system;
- the average time $\bar{\vartheta}_a$ when the last customer leaves the system;
- the probability that all customers are still in the system at $t_1 = 2.5$ min;
- the probability that all customers have left the system at $t_2 = 5$ min;

when $m = 5$ and $\vartheta_0 = 2$ min.

2.6.3.1　Case 1: Deterministic Service Time

This case is very straightforward. All customers leave the system exactly after 2 min from t_0, therefore:

- $\bar{\vartheta}_f = \bar{\vartheta}_a = \vartheta_0 = 2$ min
- the probability that all customers are still in the system at $t_1 = 2.5$ min is clearly 0;
- the probability that all customers have left the system at $t_2 = 5$ min is clearly 1.

2.6.3.2 Case 2: Exponential Service Time

In this case we start with an observation that we will use extensively also in the following chapters. Given that the average service time is $\bar{\vartheta}$, the average service rate per sever will be as usual $\mu = 1/\bar{\vartheta}$. When m servers work in parallel the overall average service rate μ_m will be m times that of a single server, therefore

$$\mu_m = m\mu \tag{2.79}$$

This is true in general for the average, but we do not know anything about the probability distribution of the time between service completions.

Nonetheless, we can observe that the problem is very similar to what discussed in Sect. 2.2.1.3. The servers working with exponential service times are equivalent to a set of m Poisson processes with inter-arrival time $\tau = \vartheta$. Therefore, the overall service process is analog to the superposition of m Poisson processes, and it is still a Poisson process with a total arrival rate given by the sum of the single arrival rates. Therefore, in this specific case it is a Poisson process with rate $\mu_m = m\mu$. The time between events is still exponentially distributed with average μ_m.

This is a general results: when m exponential servers are simultaneously busy, the overall service process has an average rate that is the sum of the single service rates, and the time between the completion of services is still exponential.

In the specific case we are considering here, no more than m customers arrive to the system, meaning that the server that completed its duty remains idle and does not contribute to the "service effort" anymore. Therefore, the system evolves into m different phases:

- phase m: when all m servers are active and the service rate is $\mu_m = m\mu$;
- phase $m-1$: when $m-1$ servers are active and the service rate is $\mu_{m-1} = (m-1)\mu$;
- phase $m-2$: when $m-2$ servers are active and the service rate is $\mu_{m-2} = (m-2)\mu$;
- ...
- phase 2: when 2 servers are active and the service rate is $\mu_2 = 2\mu$;
- phase 1: when only 1 server is active and the service rate is $\mu_1 = \mu$.

Therefore we can now answer to question 1:

$$\bar{\vartheta}_f = \frac{1}{m\mu} \tag{2.80}$$

and question 2

$$\bar{\vartheta}_a = \sum_{i=1}^{m} \frac{1}{i\mu} = \frac{H(m)}{\mu} \tag{2.81}$$

where $H(m)$ is the m-th harmonic number. In the specific case of $m = 5$ we obtain

$$\bar{\vartheta}_f = \frac{1}{m\mu} = \frac{2}{5} = 0.4 \text{ min} \tag{2.82}$$

and

$$\bar{\vartheta}_a = \frac{H(5)}{\mu} = 2 \cdot 2.283 = 4.56 \text{ min} \tag{2.83}$$

To answer question 3 let us consider what follows:

$$\Pr\{\text{all customers still in the system at } t_1\}$$

$$= \Pr\{\text{no customer completed the service at } t_1\}$$

$$= 1 - (1 - e^{-m\mu t_1}) = e^{-m\mu t_1} = 0.002 \tag{2.84}$$

To answer question 4 let us recall that the service time distribution for every server is given by

$$F_\vartheta(t) = 1 - e^{-\mu t} \tag{2.85}$$

therefore, given that the servers are independent, the probability that all of them completed the service at time t_2 is given by

$$\Pr\{\text{all service completed at } t_2\} = [F_\vartheta(t_2)]^5 = (1 - e^{-\mu t_2})^5 = 0.65 \tag{2.86}$$

Exercises

1. Voice calls arrive at a central office as a Poisson process with average arrival rate $\lambda = 20$ calls/min. Find:

   (a) the average number of calls arriving in 2 min;
   (b) the probability that 10 calls arrive in 1 min;
   (c) the probability that 40 calls arrive in 1 min.

2. Prove that in a Poisson process with average arrival rate λ the average number of requests in a given period T is equal to the variance of the number of requests in T.

3. The packet arrival process at a router interface is observed through a protocol analyzer. After an observation period long enough to reach statistical equilibrium, the measured inter-arrival time is found to be reasonably approximated as a negative exponential distribution with average $\bar{\tau} = 2$ ms. The measured average transmission time of the packets is $\bar{\vartheta} = 1.6$ ms Find:

(a) the average number of packets arriving within the average transmission time of a packet;

(b) the probability that 1 packet arrives within the average transmission time of a packet;

(c) the probability that 5 packets arrive within the average transmission time of a packet.

4. Telephone calls arrive at a PABX from three different buildings. From each building the call arrival process can be approximated as a Poisson process with average arrival rates $\lambda_1 = 10$ calls/min, $\lambda_2 = 8$ calls/min, and $\lambda_3 = 5$ calls/min, respectively. Find the average inter-arrival time at the PABX between two calls from any building, as well as its probability distribution function.

5. Telephone calls incoming to a company's customer service follow a Poisson arrival process with average rate $\lambda = 12$ calls/min. The calls are forwarded to three groups of operators based on the specific nature of the help request made by the customer, who selects the service support they need among the available options at the beginning of the call. Half of the customers select the first group of operators, whereas the rest of the calls are forwarded to the other two groups of operators in the proportion of 40% to 60%. Find the probability to receive 5 calls in 1 min for each of the three groups of operators.

6. Find the expression of the variance of a random variable with a negative exponential distribution and average $\bar{\vartheta} = \frac{1}{\mu}$.

7. Find the expression of the variance of a random variable with a uniform distribution between a minimum value ϑ_m and a maximum value ϑ_M.

8. Find the expression of the variance of a random variable with an Erlang distribution of grade r and average $\bar{\vartheta} = \frac{1}{\mu} = r\bar{\vartheta}_0$.

9. Find the expressions of the mean value and variance of a random variable with a Pareto distribution, where $t_0 > 0$ and $\alpha > 0$ are the scale parameter and shape parameter, respectively.

10. Find the average residual time of a service with an Erlang distribution of grade r and average $\bar{\vartheta} = \frac{1}{\mu} = r\bar{\vartheta}_0$.

11. Find the average residual time of a service with a Pareto distribution, where $t_0 > 0$ and $\alpha > 0$ are the scale parameter and shape parameter, respectively.

Chapter 3
Formalizing the Queuing System: State Diagrams and Birth–Death Processes

Abstract In this chapter the reader will be shown how the arrival process can be linked to the statistics of the service time, in order to achieve a model for the queuing system as a whole. The Markov Chain is the formal tool that can help solving this sort of problems in general. Here we will focus on a specific subset of Markov Chains, the so-called birth–death processes, which well match with the memoryless property of the Poisson process and of the negative exponential distribution. The general model described in this chapter will be re-used in the remainder of the book to characterize the specific queuing systems suitable for different application scenarios.

Keywords Congestion · PASTA property · Markov Chain · Birth–death process · Birth rate · Death rate · Steady state probabilities

3.1 Stateful and Time Dependent Systems

In this chapter we will introduce the mathematical modeling of relevant queuing systems suitable for the analysis of telecommunication networks. As already discussed, we are considering systems that exhibit a dynamic behavior that is the result of some underlying random process. We assume that such dynamic behavior can be described by means of a suitable set of *states* which are countable and can be associated with a *state variable*. The state variable may be a scalar that will be called k in the following, or a tuple of values represented by a vector $\mathbf{k}$.[1] Very often k represents the number of customers in the queuing system, but this is not mandatory as we will see in a few examples.

The goal of the modeling effort is to find a way to link the dynamic behavior of the state variable to the underlying random processes (i.e., arrivals, departures,

---

[1] Unless otherwise specified, in this book we will refer to systems that can be described with a scalar state variable k, which can be assumed, without loss of generality, to take values in the natural number set $k \in \mathbb{N}$. Usually $k = 0, 1, \ldots, n, \ldots$.

© Springer Nature Switzerland AG 2023
F. Callegati et al., *Traffic Engineering*, Textbooks in Telecommunication Engineering, https://doi.org/10.1007/978-3-031-09589-4_3

service times, queuing policies, etc.) and to the performance metrics that are of interest for engineering the system. This is mostly done under the assumption of random processes that are ergodic. This is a basic assumption that makes system engineering possible. If the system is ergodic, the statistical quantities that can be calculated from a series of samples either obtained in time from a given system or obtained at the same time from many systems that behave in the same way. This makes it possible to infer the future behavior of a system from the observation and analysis of the past behavior, as well as to infer the behavior of a system from the observation of the behavior of another similar system.

It is worth stating here that we are interested into three possible engineering actions:

analysis: given a quantitative characterization of the random processes and the dimensioning of the system, find relevant performance metrics;

design: given some target values for the performance metrics, find the related dimensioning of the system (queue size, number and capacity of the servers, etc.);

planning: link the system dimensioning to a given evolution in time of the quantitative characteristics of the underlying random process (i.e., provide a time variant system design).

The mathematical tool typically used to perform the target analysis of a queuing system is the Markov Chain (MC), which is briefly introduced in Appendix A. In this chapter we will introduce and focus on a specific subset of ergodic MCs that are simpler to characterize and, at the same time, very relevant for the specific problems we have to deal with, the so-called *birth–death processes*.

3.2 Defining Congestion as a Sample State

Before delving into the details of the analysis of a birth–death process, it is very important to discuss and better understand some issues related to the system performance metrics. As a matter of fact, an effective analysis can be carried out only with a clear understanding of the performance metrics, their meaning and definition. So let us start by defining the basic concept of congestion, i.e., the condition of the system causing some drop in the quality of service perceived by the customer.

When a customer requests a service and the request cannot be immediately satisfied by a server, we say that there is a *congestion* or *blocking* condition. Depending on the characteristics of the system, the consequence of congestion may be that the customer cannot be served at all and is *lost*, or the service is *delayed* and the customer is put on hold.

Performance analysis of a queuing system almost always implies that we can properly define the conditions leading to congestion, and that we can calculate the probability that a loss or delay happens. As a matter of fact, if we know how to compute the probability that a customer is lost or delayed in an ergodic system, then we also know the percentage of customers that are lost or delayed, which very often is the target design parameter.

If we know how to calculate the steady state probabilities of a system and how to link congestion to one or more system states, it should be possible to derive the congestion probabilities from the steady state probabilities. To better understand the latter point, it is necessary to look for a more formal and precise definition of congestion and of the probability that it happens. Generally in the literature we find two possible alternatives:

time congestion: define the congestion probability as the probability that the queuing system is in a state that may lead to congestion (e.g., when all servers are occupied);

call congestion: define the congestion probability as the probability that the queuing system is in a state that will lead to congestion when (i.e., conditioned to the fact that) a new service request happens.

Assuming that it is possible to say which are the states that may lead to congestion, and this is usually easy to do at least in the more common queuing systems, the time congestion is just the sum of the steady state probabilities of such states and is therefore rather easy to calculate. This is not true for the call congestion.

At the same time, it is easy to understand that the latter and not the former is the most meaningful performance metric. This is easily shown with a simple example. Let us assume to have a queuing system with m servers and no queue, serving a population of m users. Let us assume the system is ergodic and we can calculate the steady state probabilities of this system $\pi_k \; \forall k$. Obviously, $\pi_k \neq 0$ if and only if $k \leq m$, whereas $\pi_k = 0 \; \forall k > m$. For this system the time congestion is given by the fact that the system is in state m, i.e., no server is free and a new request will be lost, if any arrives. Therefore the time congestion probability is given by π_m. However, this is not a meaningful quantity because, being m the total number of customers, it is rather obvious that no customer will ever arrive when the system is in state m, and therefore the call congestion probability is 0.

In general the time congestion is easier to calculate but provides just an approximate estimate of the real congestion probability, while the call congestion provides a more meaningful evaluation, i.e., the correct assessment of the quality perceived by the customers, but may be more difficult if not impossible to calculate with a closed-form formula. This is something we have to cope with, especially when dealing with complex queuing systems.

3.2.1   The PASTA Property

In spite of what discussed above, when the arrival process is a Poisson process something happens that changes the rule and the call congestion probability is actually equal to the time congestion probability. As we will show in the following, this is linked to the fact that the Poisson process has its own dynamics and does not depend on the state of the system.

This property is usually called PASTA, from the acronym of the statement that "Poisson Arrivals See Time Averages." This is one possible way to say that the call congestion probability is equal to the time congestion probability.

It is not difficult to prove the PASTA property, considering an ergodic MC subject to arrivals distributed according to a Poisson process. Let us consider that the arrival process is independent of the state of the system, since the Poisson arrival rate is a constant that does not change in time. Therefore:

$$\text{Pr\{customer arrival } | \text{ time congestion\}} = \text{Pr\{customer arrival \}} \tag{3.1}$$

Now let us recall, from the definition of conditional probability, that we can write:

$$\Pr\{A \mid B\}\Pr\{B\} = \Pr\{B \mid A\}\Pr\{A\} = \Pr\{A, B\} \tag{3.2}$$

Therefore:

$$\text{Pr\{time congestion } | \text{ customer arrival \}} \cdot \text{Pr\{customer arrival \}}$$
$$= \text{Pr\{customer arrival } | \text{ time congestion\}} \cdot \text{Pr\{time congestion\}}$$
$$= \text{Pr\{customer arrival \}} \cdot \text{Pr\{time congestion\}} \tag{3.3}$$

and:

$$\text{Pr\{time congestion } | \text{ customer arrival \}} = \text{Pr\{time congestion\}} \tag{3.4}$$

and finally, considering the definition of call congestion we gave above, we obtain the fundamental result we were looking for:

$$\text{Pr\{call congestion\}} = \text{Pr\{time congestion\}} \tag{3.5}$$

The PASTA property is a very important result that we will extensively use in the following. It brings the problem to study the congestion in a queuing system down to the problem of calculating the steady state probabilities and properly identifying which states may lead to a congestion condition.

3.3 Birth–Death Processes

Now let us consider a specific kind of MC: the *birth–death process* (or BD process).

A *birth–death process* is a continuous time MC that allows state transitions only between adjacent state values. Therefore, the transition rates are $q_{i,j} = 0$ if $|i - j| > 1$.

As for general MCs, the BD process may be described with a state diagram, as illustrated in Fig. 3.1 for a generic BD process. For the sake of simplicity and to distinguish between BD processes and generic MCs, in the remainder of this book we will use the following notation in case of BD processes:

- $\lambda_k = q_{k(k+1)}$: transition rate from state k to state $k + 1$
- $\mu_k = q_{k(k-1)}$: transition rate from state k to state $k - 1$
- $P_k(t) = \pi_k(t)$: steady state probability of state k

λ_k and μ_k are called *birth rate* and *death rate* in state k, respectively.

With these definitions, the transition rate matrix $\mathscr{Q}$ for a BD process becomes the following tridiagonal matrix:

$$\mathscr{Q} = \begin{bmatrix} -\lambda_0 & \lambda_0 & 0 & 0 & \cdots \\ \mu_1 & -(\mu_1 + \lambda_1) & \lambda_1 & 0 & \cdots \\ 0 & \mu_2 & -(\mu_2 + \lambda_2) & \lambda_2 & \cdots \\ \vdots & \vdots & & \ddots & \ddots \ddots \end{bmatrix} \quad (3.6)$$

and Eq. (A.32) can be simplified as:

$$\frac{d}{dt} P_0(t) = -\lambda_0 P_0(t) + \mu_1 P_1(t)$$

$$\frac{d}{dt} P_k(t) = \lambda_{k-1} P_{k-1}(t) - (\lambda_k + \mu_k) P_k(t) + \mu_{k+1} P_{k+1}(t) \quad \forall k > 0$$

$$(3.7)$$

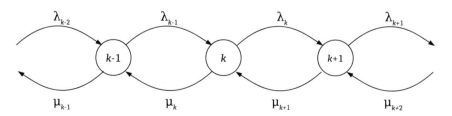

Fig. 3.1 State diagram of a generic birth–death process

For an ergodic BD process, Eq. (3.7) simplifies to:

$$
\begin{cases}
0 = -\lambda_0 P_0 + \mu_1 P_1 \\
0 = \lambda_0 P_0 - (\lambda_1 + \mu_1) P_1 + \mu_2 P_2 \\
0 = \lambda_1 P_1 - (\lambda_2 + \mu_2) P_2 + \mu_3 P_3 \\
\vdots
\end{cases}
\tag{3.8}
$$

which can be solved iteratively to obtain the following:

$$
\begin{cases}
P_1 = \frac{\lambda_0}{\mu_1} P_0 \\
P_2 = \frac{\lambda_1}{\mu_2} P_1 = \frac{\lambda_1}{\mu_2} \frac{\lambda_0}{\mu_1} P_0 \\
P_3 = \frac{\lambda_2}{\mu_3} \frac{\lambda_1}{\mu_2} \frac{\lambda_0}{\mu_1} P_0 \\
\vdots \\
P_k = \prod_{i=0}^{k-1} \frac{\lambda_i}{\mu_{i+1}} P_0 \quad k > 0
\end{cases}
\tag{3.9}
$$

This result is not surprising. We already know that the steady state probabilities of an ergodic MC depend on an arbitrary constant, which we conventionally choose by imposing that the sum of the steady state probabilities is 1. This constant is P_0 in the case of BD processes. Therefore:

$$
P_0 = \frac{1}{1 + \sum_{\forall k > 0} \prod_{i=0}^{k-1} \frac{\lambda_i}{\mu_{i+1}}}
\tag{3.10}
$$

and, in conclusion:

$$
P_k = \frac{\prod_{i=0}^{k-1} \frac{\lambda_i}{\mu_{i+1}}}{1 + \sum_{\forall k > 0} \prod_{i=0}^{k-1} \frac{\lambda_i}{\mu_{i+1}}} \quad k > 0
\tag{3.11}
$$

It is possible to prove that if $P_0 > 0$ the BD process is ergodic. We do not provide this formal proof here, but it is rather intuitive that, since all P_k depend on P_0, they could not exist if P_0 would not have a finite value.

Finally, from Eqs. (A.38) and (A.39), the probability distribution of the time τ_k spent in state k is:

$$
\Pr\{\tau_k < t\} = 1 - e^{-(\lambda_k + \mu_k)t}
\tag{3.12}
$$

3.4 Queuing Systems, Memoryless Property, and BD Processes

At this stage we have elaborated on some important tools in the direction of designing a general framework to describe a subset of queuing systems that may be used to model a teletraffic system, and the related problems we are interested to solve. In the previous chapter we discussed the analytical tools used to describe random arrivals and service times. In the previous section we have briefly described the BD process, which is useful to describe the statistics of the state of a system that exhibits some specific characteristics. Finally, proving the PASTA property we have shown that, in case of Poisson arrivals and ergodic systems, the steady state probabilities can be used to calculate performance metrics.

What we are still missing is a complete link between the arrival and service processes, and the system state description, if it exists. In this section we will elaborate on this topic. As usual, the goal is to provide not a formal mathematical proof, but rather some intuitive evidence of such a link.

Let us imagine that we want to find the MC underlying an ergodic queuing system which is visited by customers arriving according to a Poisson process with rate λ. The Poisson process is independent of the state of the system, therefore λ does not change in time, regardless of what happens to the queuing system. We discussed in Sect. 2.2.1 that, in the case of a Poisson process, arrivals are never overlapping, therefore customers will arrive randomly at the system, but strictly one at a time. If the state k represents the number of customers in the system, then k will increase with arrivals and decrease with departures. If the arrivals can be described with a Poisson process with rate λ, then it immediately follows that, if k increases, it can only become $k + 1$, and the transition rate from k to $k + 1$ is given by λ, i.e., the arrival rate of the Poisson process. This is true for any k.

Now let us assume that the service time is exponentially distributed with average $\bar{\vartheta} = 1/\mu$ for any server.[2] Following what discussed in Exercise 2.6.3, if m_k is the number of busy servers in state k, then the service rate in state k is $m_k \mu$, and the time between the completion of two consecutive services is exponentially distributed with average $\bar{\vartheta}/m_k$. Again, no server will complete the service at the same time as another server. As above, if k decreases, it can only become $k - 1$, and the transition rate is $m_k \mu$, i.e., the departure rate from the system.

As a matter of fact, in both cases we face memoryless processes. Therefore, the evolution of the system from a given state does not depend on the past states, but only on the current state and on the time elapsed since we started observing it, which determine the subset of future states, i.e., either $k-1$ or $k+1$. Finally, let us consider arrivals and departures as a whole. These are two processes with memoryless exponential inter-arrival times, the former with rate λ, the latter with rate $m_k \mu$. If

---

[2]Recall that the servers are identical from the customer perspective, therefore they all behave the same.

we imagine to combine them to see when the next event will occur (either arrival or departure), the result is a new memoryless process with exponential inter-event times, with rate $\lambda + m_k \mu$, according to what already discussed in Sect. 2.2.1.3 and in Exercise 2.6.3. This is to say that the time the system will spend in state k is exponentially distributed with average

$$\bar{\tau}_k = \frac{1}{\lambda + m_k \mu}$$

which is exactly in line with Eq. (A.39) in Appendix A.

In practice, the discussion above shows that any queuing system with Poisson arrivals and exponential service times has the properties of a MC, with the additional feature that the state transitions are limited between adjacent states. Therefore, it can be modeled as a BD process and the steady state probabilities can be calculated with the formulas introduced above, in particular using Eq. (3.11).

3.5 Examples and Case Studies

3.5.1 The Poisson Process as a Birth-Only Process

The Poisson process is very closely related to the MC, as we have seen in the discussion in Sect. 3.4. Let us imagine a BD process that has a constant birth rate $\lambda_k = \lambda \ \forall k$ and a null death rate $\mu_k = 0 \ \forall k$. Obviously, this system cannot be stationary since the state evolves in time by increasing in value in a never-ending transient to infinity. Therefore, it cannot be studied using formula (3.9), but we need to refer to its time dependent version as in formula (3.7). We then obtain:

$$\frac{d}{dt} P_0(t) = -\lambda P_0(t)$$
$$\frac{d}{dt} P_k(t) = \lambda P_{k-1}(t) - \lambda P_k(t) \quad \forall k > 0$$

(3.13)

This set of differential equations has the following starting conditions:

$$\begin{cases} P_0(0) = 1 \\ P_k(0) = 0 \quad \forall k > 0 \end{cases}$$

(3.14)

and can be easily solved by substitution, finding the following solution:

$$\begin{cases} P_0(t) = e^{-\lambda t} \\ P_k(t) = \dfrac{(\lambda t)^k}{k!} e^{-\lambda t} \quad k > 0 \end{cases}$$

(3.15)

$P_k(t)$ is the probability that the process sends k requests in the observed interval $[0 : t]$. If this interval is called T, then $P_k(T) = \dfrac{(\lambda T)^k}{k!} e^{-\lambda T}$ is equal to the Poisson formula $P(k, T)$ in (2.12).

It is worth mentioning that following (3.12) it is easy to show that the probability distribution of the time spent in state k is:

$$F_{\tau_k}(t) = 1 - e^{-\lambda t} \tag{3.16}$$

for all values of k. This is to say that the time τ between two arbitrarily chosen births in this process has an exponential distribution with parameter λ:

$$F_\tau(t) = 1 - e^{-\lambda t} \tag{3.17}$$

which is what already shown for the inter-arrival time of the Poisson process in Sect. 2.2.1.2.

This special BD process with only births and no deaths is simply a Poisson process, thus closing the link between the Poisson process, the memoryless property of the exponential distribution, and the fundamental property of MCs.

3.5.2 Alarm Reporting

In a company, a single operator receives calls from different business units reporting failures, problems, and other alarms. When the employees call the operator, they can choose one of the two numbers devoted to this service:

- number 100 is dedicated to high priority calls (let us call them type A calls) that are used to report events that require immediate attention;
- number 200 is dedicated to low priority calls (let us call them type B calls) that are used to report relevant events, but not as urgent as the previous ones.

The local Private Automatic Branch eXchange (PABX) used by the company does not have call waiting capabilities and therefore behaves as follows:

- when a type B call arrives and the operator is busy answering either a type A or a type B call, the incoming call is blocked;
- when a type A call arrives and the operator is busy answering another type A call, the incoming call is blocked;
- when a type A call arrives and the operator is busy answering a type B call, the type B call is *pre-empted*, i.e., it is interrupted and the operator immediately answers the incoming type A call.

Calls arrive according to two Poisson processes with average arrival rates $\lambda_A = 3$ calls/h $= 0.05$ calls/min and $\lambda_B = 6$ calls/h $= 0.1$ calls/min, and the call duration can be assumed exponentially distributed with averages $\bar{\vartheta}_A = 0.5$ min and $\bar{\vartheta}_B = 1$ min for call types A and B, respectively. Therefore, $\mu_A = 2$ calls/min and $\mu_B = 1$ call/min.

The system performance metrics of interest to be calculated are

1. the blocking probability of type A calls π_{pA};
2. the blocking probability of type B calls π_{pB};
3. the probability π_s that a B call is dropped after being answered;
4. the overall probability π_i that a type B call cannot be completed either because it is blocked or because it is dropped.

At first, let us find the offered traffic:

- for type A calls the offered traffic is $A_{0A} = \lambda_A \bar{\vartheta}_A = 0.025$ E
- for type B calls the offered traffic is $A_{0B} = \lambda_B \bar{\vartheta}_B = 0.1$ E

The queuing system can be described as a MC with three states:

- 0: no active calls and operator idle;
- A: the operator is answering a type A call;
- B: the operator is answering a type B call.

The state diagram of the system is as shown in Fig. 3.2: both states A and B are reachable from state 0, and state 0 is reachable from both states A and B. On the other hand, state A can be reached from state B but state B cannot be reached from state A, because type A calls are allowed to interrupt and pre-empt type B calls, whereas the opposite is not allowed. The state transition rates are as shown in the figure.

The Chapman–Kolmogorov equations in this case can be written as:

$$\begin{cases} \pi_0(\lambda_A + \lambda_B) = \pi_A \mu_a + \pi_B \mu_B \\ \pi_B(\lambda_A + \mu_B) = \pi_0 \lambda_B \\ \pi_A \mu_a = \lambda_A(\pi_0 + \pi_B) \end{cases} \qquad (3.18)$$

together with:

$$\pi_0 + \pi_A + \pi_B = 1 \qquad (3.19)$$

Fig. 3.2 Markov chain state diagram and transition rates of the alarm reporting use case

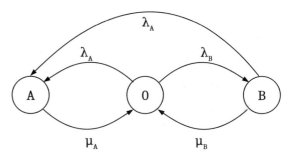

It is rather easy to calculate the steady state probabilities:

$$\begin{cases} \pi_0 = \left[1 + \frac{\lambda_B}{\lambda_A + \mu_B} + \frac{\lambda_A}{\mu_A}\left(1 + \frac{\lambda_B}{\lambda_A + \mu_B}\right)\right]^{-1} = \left[\left(1 + \frac{\lambda_B}{\lambda_A + \mu_B}\right)\left(1 + \frac{\lambda_A}{\mu_A}\right)\right]^{-1} = 0.891 \\[2mm] \pi_A = \pi_0 \frac{\lambda_A}{\mu_A}\left(1 + \frac{\lambda_B}{\lambda_A + \mu_B}\right) = \frac{\lambda_A}{\lambda_A + \mu_A} = 0.024 \\[2mm] \pi_B = \pi_0 \frac{\lambda_B}{\lambda_A + \mu_B} = 0.085 \end{cases}$$

$$(3.20)$$

Then, by applying the PASTA property we can obtain the requested performance metrics. The blocking probability of type A calls is simply given by the probability that the system is in state A:

$$\pi_{pA} = \pi_A = \frac{\lambda_A}{\lambda_A + \mu_A} = \frac{A_{0A}}{1 + A_{0A}} = 0.024 \tag{3.21}$$

The expression of π_{pA} corresponds to the value of Erlang B formula $\mathscr{B}(1, A_{0A})$, which gives the blocking probability of a single-server system without queuing space subject to traffic A_{0A}, as will be discussed in detail in Chap. 4. This is in line with what suggested by intuition. Calls of type A are not affected in any way by calls of type B, therefore they see a pure loss system with only one server.

The blocking probability of type B calls is simply given by the probability that the system is in state A or B, i.e., $1 - \pi_0$:

$$\pi_{pB} = 1 - \pi_0 = 0.11 \tag{3.22}$$

A type B call is completed if and only if no type A calls arrive during its service time. Therefore, we can use the Poisson formula:

$$\Pr\{0 \text{ arrivals of type } A \text{ in } \vartheta_B\} = P_A(0, \vartheta_B) = e^{-\lambda_A \vartheta_B} \tag{3.23}$$

and by averaging over all possible values of ϑ_B:

$$\pi_s = \int_0^\infty (1 - P_A(0, t))\, f_{\vartheta_B}(t)dt = \int_0^\infty (1 - e^{-\lambda_A t})\mu_B e^{-\mu_B t} dt = \frac{\lambda_A}{\lambda_A + \mu_B} = 0.0476 \tag{3.24}$$

Finally, a type B call cannot be completed if it is blocked or dropped after being accepted. Therefore π_i is given by two possible disjoint events: the call is blocked at arrival, the call is not blocked at arrival but is pre-empted. The probability of the former is known already as π_{pB}; the probability of the latter can be calculated as the joint probability of being accepted and then dropped by pre-emption. These two events are independent and therefore the joint probability is the product of the probabilities of the two events. In summary:

$$\pi_i = \Pr\{\text{type } B \text{ blocked}\} + \Pr\{\text{type } B \text{ not blocked}\} \cdot \Pr\{\text{type } B \text{ pre-empted}\}$$
$$= \pi_{pB} + [(1 - \pi_{pB})\pi_s] = \pi_{pB} + \pi_0\pi_s = 0.11 + 0.04 = 0.15 \tag{3.25}$$

3.5.3 Taxis at the Airport

Let us now analyze a sample case that does not refer to telecommunications but can be modeled as a BD process. Let us consider a typical taxi service located outside the arrival hall of an airport. The taxi parking area is limited and only up to N taxis can be parked waiting for customers. On the other hand, the area where customers wait in case there are no taxis available is very large, so large that we can assume that the number of waiting customers is not upper-bounded.

Let us assume that the customers arrive according to a Poisson process with rate λ_c, and that the taxis arrive at the parking area also according to a Poisson process with rate λ_t. We are interested in finding:

- the average number of taxis $\bar{n}_t$ waiting for customers in the parking area;
- the average number of customers $\bar{n}_c$ waiting in the queue;
- the probability π_{wt} that a taxi has to wait for a customer in the parking area;
- the probability π_{bt} that a taxi arrives at the parking area and cannot park because the parking area is full;
- the probability π_{wc} that a customer has to wait for a taxi.

In this case, defining the state variable simply as the number of customers in the system is not appropriate, since it would not allow to capture the behavior of the system when there are no customers and taxis queue up in the parking area. Therefore, we extend the definition of state variable k as follows:

- $k \in [1, \infty]$ if k customers are waiting in the queue because no taxis are available;
- $k = 0$ if no customers are waiting in the queue and no taxis are waiting in the parking area;
- $k \in [-N, -1]$ if $n_t = -k$ taxis are parked in the parking area.

Accordingly, k increases to $k + 1$ whenever a new customer arrives, $\forall k \in [-N, \infty]$, and decreases to $k - 1$ whenever a new taxi arrives, $\forall k \in [-N + 1, \infty]$.

The random processes that determine the dynamics of the system, i.e., customer and taxi arrivals, are both memoryless, and transitions may happen only between adjacent states. Therefore, the steady state probabilities can be obtained with the formulas of a BD process (3.11), with the following values for the transition rates:

$$\lambda_k = \lambda_c \quad \forall k \in [-N, \infty]$$
$$\mu_k = \lambda_t \quad \forall k \in [-N + 1, \infty] \tag{3.26}$$

However, it must be noted that in this case the state with the smallest value of k is P_{-N} and not P_0. Therefore, the index used in the state probabilities (3.11) must be shifted left by N units, resulting in:

$$P_k = P_{-N} \left(\frac{\lambda_c}{\lambda_t}\right)^{N+k}$$

$$P_{-N} = \frac{1}{\sum_{k=-N}^{\infty} \left(\frac{\lambda_c}{\lambda_t}\right)^{N+k}} = \frac{1}{\sum_{h=0}^{\infty} \left(\frac{\lambda_c}{\lambda_t}\right)^{h}} = 1 - \frac{\lambda_c}{\lambda_t} \tag{3.27}$$

It is important to highlight that the state probabilities in (3.27) exist only if P_{-N} exists, i.e., for $\lambda_c < \lambda_t$, otherwise the geometric series on which P_{-N} depends does not converge. This condition has an intuitive meaning: if the customer arrival rate λ_c is larger than the taxi arrival rate λ_t, then in the long term there will not be enough taxis available to pick up all customers, and the number of customers waiting in the queue increases to infinity. In other words, the condition $\lambda_c < \lambda_t$ ensures that the BD process is ergodic.

Now that we know the steady state probabilities, we can obtain the system performance metrics as requested. The *average number of taxis waiting for customers in the parking area* is given by:

$$\bar{n}_t = \sum_{k=-N}^{-1} (-k) P_k = \left(1 - \frac{\lambda_c}{\lambda_t}\right) \sum_{k=-N}^{-1} (-k) \left(\frac{\lambda_c}{\lambda_t}\right)^{N+k}$$

$$= \left(1 - \frac{\lambda_c}{\lambda_t}\right) \sum_{h=0}^{N-1} (N-h) \left(\frac{\lambda_c}{\lambda_t}\right)^{h} = N - \frac{\lambda_c}{\lambda_t} \frac{1 - \left(\frac{\lambda_c}{\lambda_t}\right)^{N}}{1 - \frac{\lambda_c}{\lambda_t}} \tag{3.28}$$

The *average number of customers waiting in the queue* is given by:

$$\bar{n}_c = \sum_{k=1}^{\infty} k P_k = \left(1 - \frac{\lambda_c}{\lambda_t}\right) \sum_{k=1}^{\infty} k \left(\frac{\lambda_c}{\lambda_t}\right)^{N+k}$$

$$= \left(1 - \frac{\lambda_c}{\lambda_t}\right) \left(\frac{\lambda_c}{\lambda_t}\right)^{N} \sum_{k=1}^{\infty} k \left(\frac{\lambda_c}{\lambda_t}\right)^{k} = \frac{\left(\frac{\lambda_c}{\lambda_t}\right)^{N+1}}{1 - \frac{\lambda_c}{\lambda_t}} \tag{3.29}$$

Since the arrival processes of customers and taxis are Poisson processes, the PASTA property holds and we can easily find the requested congestion probabilities.

The probability that a taxi has to wait for a customer in the parking area is the sum of the steady state probabilities of all states that do not see any customer waiting in the queue, since a taxi waits when there are no customers waiting. Therefore:

$$\pi_{wt} = \sum_{k=-N+1}^{0} P_k = \left(1 - \frac{\lambda_c}{\lambda_t}\right) \sum_{k=-N+1}^{0} \left(\frac{\lambda_c}{\lambda_t}\right)^{N+k}$$

$$= \left(1 - \frac{\lambda_c}{\lambda_t}\right) \frac{\lambda_c}{\lambda_t} \sum_{h=0}^{N-1} \left(\frac{\lambda_c}{\lambda_t}\right)^{h} = \frac{\lambda_c}{\lambda_t} - \left(\frac{\lambda_c}{\lambda_t}\right)^{N+1} \tag{3.30}$$

The probability that a taxi arrives at the parking area and cannot park because the parking area is full is simply given by:

$$\pi_{bt} = P_{-N} = 1 - \frac{\lambda_c}{\lambda_t} \tag{3.31}$$

The probability that a customer has to wait for a taxi is the sum of the steady state probabilities of all states that do not see any taxi in the parking area, therefore when a customer arrives they have to wait for the next taxi:

$$\pi_{wc} = \sum_{k=0}^{\infty} P_k = \left(1 - \frac{\lambda_c}{\lambda_t}\right) \sum_{k=0}^{\infty} \left(\frac{\lambda_c}{\lambda_t}\right)^{N+k}$$

$$= \left(1 - \frac{\lambda_c}{\lambda_t}\right) \left(\frac{\lambda_c}{\lambda_t}\right)^{N} \sum_{k=0}^{\infty} \left(\frac{\lambda_c}{\lambda_t}\right)^{k} = \left(\frac{\lambda_c}{\lambda_t}\right)^{N} \tag{3.32}$$

As a final note, let us consider the two extreme cases when $\lambda_c = 0$, i.e., when no customers arrive, and when $\lambda_c \to \lambda_t$, i.e., when the customer arrival rate is very close to the taxi arrival rate. When $\lambda_c = 0$, the parking area is always full ($\bar{n}_t = N$), there are no customers waiting ($\bar{n}_c = 0$ and $\pi_{wc} = 0$), and any newly arriving taxi finds the parking area full ($\pi_{wt} = 0$ and $\pi_{bt} = 1$). On the other hand, when $\lambda_c \to \lambda_t$, there are no taxis waiting in the parking area ($\bar{n}_t \to 0$ and $\pi_{wt} \to 0$), no taxi finds the parking area full ($\pi_{bt} \to 0$), any newly arriving customer waits ($\pi_{wc} \to 1$), and the customer queue becomes very large ($\bar{n}_c \to \infty$).

Exercises

1. A processing system is equipped with two processors and is subject to two types of processing tasks: *monolithic* tasks that must be executed entirely by a single processor, *modular* tasks that must be split in two sub-tasks to be executed by the two processors in parallel. As the first approximation, we assume that the processing system does not have any queuing space to store tasks waiting to be processes, so it behaves as a loss system. Specifically, monolithic tasks are served if at least one of the two processors is free, whereas modular tasks are served only if both processors are free, otherwise they are rejected. Assume that the arrival processes of the two task types follow a Poisson distribution, with rates $\lambda_1 = 1000$

tasks/s for monolithic tasks and $\lambda_2 = 500$ tasks/s for modular tasks. Assume also that the processing time of a monolithic task and of each sub-task of a modular task is exponentially distributed with average $\bar{\vartheta} = 500 \ \mu s$.

(a) Represent the system described above by means of a Markov Chain.
(b) Find the state probabilities at the equilibrium.
(c) Find the rejection probability for monolithic tasks.
(d) Find the rejection probability for modular tasks.
(e) Compute the total traffic offered to the system as well as the throughput.

2. Consider a queuing system where an operator provides information to customers. Suppose that customers arrive at the operator's desk according to a Poisson process with frequency $\lambda = 120$ customers/hour. However, when new customers find other customers already waiting in the queue, they are discouraged to queue up and may give up waiting. The longer the queue, the higher the probability of giving up by new customers. If k represents the number of customers in the system, we can assume that the arrival rate of customers who actually join the queue can be modeled as follows:

$$\lambda_0 = \lambda \text{ when } k = 0$$

$$\lambda_k = \frac{\lambda}{k} \text{ when } k > 0$$

Assuming that the service time follows a negative exponential distribution with mean value $\bar{\vartheta} = 1$ minute, find:

(a) the state probabilities P_k of the system at the equilibrium;
(c) the traffic A_0 actually offered to the system, the throughput A_s, and the traffic lost A_p due to customers giving up;
(c) the average number of customers in the system $\bar{k}$;
(d) the average waiting time $\bar{t}_w$ for a customer who arrives and decides to join the queue.

3. A company is located on two sites, both equipped with a Private Automatic Branch Exchange (PABX) for internal call routing between them. The interconnection between the two sites is implemented through $m = 4$ dedicated telephone lines, rented from an operator. Traffic between the two sites is generated according to a Poisson process and can be classified into two types:

- type A traffic, with an average call arrival rate $\lambda_A = 4$ calls/hour and an average call duration $\bar{\vartheta}_A = 3$ min;
- type B traffic, with an average call arrival rate $\lambda_B = 16$ calls/hour and an average call duration $\bar{\vartheta}_B = 3$ min.

Type A traffic is to be considered as a premium service and must therefore be given higher priority than type B traffic. This can be achieved by applying a *trunk reservation* policy that works as follows:

- as long as the number of busy lines (which is equal to the number of calls currently in progress between the two sites) is less than N (to be determined), calls of both types A and B are accepted;
- when the number of busy lines is equal to or greater than N, only calls of type A are accepted.

(a) Draw the state diagram that represents the behavior of the trunk reservation system for generic values of N and m.
(b) Find the expression of the steady state probabilities as a function of the probability P_0 of the system being empty, for any N and m.
(c) In the case $m = 4$, find the value of N for which the blocking probability for type A traffic is $\pi_A \leq 10^{-3}$, and compute the corresponding blocking probability π_B for type B traffic.

4. The output interface of an IoT device is connected to two lines in parallel, line A and line B, with capacity $C_A = 256$ Kbit/s and $C_B = 64$ Kbit/s, respectively. For line access management, the interface is equipped with a small queue whose capacity is limited to $L = 2$ packets, in order to minimize the queuing delay. The interface is configured in such a way that:

- when a new packet arrives and both lines are free, the packet is always transmitted on the fastest line;
- when a new packet arrives and one of the lines is busy, the packet is transmitted on the line available;
- when a new packet arrives and both lines are busy, the packet is queued, or dropped if the queue is full.

Assume that the packets have a random length with exponential distribution and average value $D = 500$ bytes, and that they arrive at the interface according to a Poisson process with an average arrival rate $\lambda = 64$ packets/s.

(a) Draw the state diagram that represents the behavior of the queuing system considered above, determining the transition frequencies between the states.
(b) Find the steady state probabilities.
(c) Find the probability that a packet must wait in the queue and the blocking probability.

5. A company is located on two sites, both equipped with a PABX. The two PABXs are interconnected by means of $m = 2$ dedicated lines. The total traffic between the two sites is Poissonian and is estimated to arrive at rate $\lambda = 20$ calls/hour. The random call holding time is exponentially distributed with mean value $\bar{\vartheta} = 3$ min. At first, the PABX does not provide any priority mechanism, so all calls are treated in the same way and there is no waiting space to hold calls when all lines are busy.

However, the incoming traffic actually consists of two kinds of flows, one with an average call frequency $\lambda_1 = 15$ calls/hour, the other with an average call frequency $\lambda_2 = 5$ calls/hour. The former flow should be given lower priority than the latter one. By upgrading the PABX it is possible to put calls on hold and differentiate traffic flow priority: high priority calls from flow 2 that find all the lines busy are queued (assuming a queue of infinite length) and served when a line becomes free, whereas low priority calls from flow 1 are treated as in the case without waiting space (and therefore dropped when all lines are busy).

(a) Find the blocking probability under the assumption of a system that treats all calls in the same way and without waiting space.
(b) Draw the state diagram of a Markov Chain describing the upgraded system with priority scheduling and waiting space for high priority calls.
(c) Find the steady state probabilities of the Markov Chain represented above.
(d) Find the probability of queuing high priority calls and the probability of blocking low priority calls.
(e) Find the throughput of the system with and without priority scheduling.

6. Consider a mobile telephone system in which each cell is served by a base station with 3 radio carriers, each capable of 8 telephone channels, for a total of $m = 24$ channels per cell. Each cell includes N users who generate, according to a Poisson process, on average $\lambda_u = 3$ calls/hour each, with an average duration per call of $\bar{\vartheta} = 2$ min. Since users can move from cell to cell, the system must provide for the possibility of transferring a call from one cell to another. Therefore, in a given cell call arrivals can be generated either by users already located in the cell who decide to start a new call, or by users already in a call coming from another cell after a handover procedure. Similarly, calls ongoing within a cell may end either because users hang up or because they leave the cell.

Assuming a uniform distribution of users and considering all cells equal to each other, we can assume that the average number of users located in a given cell is constant and equal to N, and that N is also the average number of users in neighbor cells who can potentially move to the considered cell. Assume then that each user moves to an adjacent cell according to a Poisson process with frequency $\lambda_h = 2$ handovers/hour.

(a) Assuming at first that user mobility can be neglected, find the maximum value of N such as to guarantee a call blocking probability $\pi_p \leq 1\%$.
(b) Draw the state diagram of a Markov Chain that describes the behavior of a given cell, considering the user mobility.
(c) Write the expression, as a function of N, of the total arrival rate λ_t and departure rate μ_t in the considered cell, as well as the offered traffic A_0 per cell.
(d) Find the maximum number of acceptable users per cell in order to have a call blocking probability $\pi_p \leq 1\%$.
(e) Due to particular events, it may happen that the user mobility is not uniform anymore, but there is a concentration towards a particular cell. Assume that

this situation creates a concentration of users leading to a 50% increase of N. Find the number of channels that are needed to maintain the same quality of service as above when such concentration events happen, taking into account that the capacity can be increased in terms of radio carriers and not individual channels.

7. The output interface of an IP router is equipped with a transmission line of capacity $C = 2.048$ Mbit/s. The interface receives packets generated by two types of services: *real time* and *best effort*. Real time traffic has stringent requirements in terms of transmission delay, whereas best effort traffic does not have particular service requirements. Let us consider a situation in which the best effort traffic is very high and the related packets are queued in a very large memory, which can be assumed to be always non-empty for the entire period of interest. In other words, the assumption is that as soon as the transmission line becomes free, it is immediately occupied by a best effort packet taken from the queue. The size of best effort packets can be assumed to be exponentially distributed with an average value of $D_b = 512$ bytes.

The real time packets are queued in a separate, very small memory, capable of storing only one packet at a time. These packets arrive according to a Poisson process with average rate $\lambda = 400$ packets/s, and their size is exponential with an average value of $D_r = 128$ bytes. Two modes of operation are possible:

- a newly arrived real time packet waits for the end of the transmission of the best effort packet currently being served (non-preemptive mode);
- when a real time packet arrives, the transmission of the best effort packet currently being served is interrupted to transmit the real time packet, and the best effort packet is dropped (preemptive mode).

(a) Draw the state diagram of a Markov Chain representing the behavior of the system operating in non-preemptive mode.
(b) Find the loss probability for real time packets and the related throughput.
(c) Repeat the two steps above for the preemptive mode.

8. A large company has deployed a modern PABX, which is connected to the public telephone network by means of a "very high" number of lines. The company employees, who generate a large amount of outbound traffic, are split into two classes:

- class B (low priority), which "sees" a loss system: when a class B call arrives, it is served only if at least one line is free, otherwise the call is blocked; the arrival process for class B calls is Poissonian with rate $\lambda_B = 50$ calls/hour;
- class A (high priority), which "sees" a system with infinite queuing space: class A calls compete for free lines with class B calls, but in case all lines are busy, class A calls are put on hold in an infinite FIFO queue; also for class A the arrival process is Poissonian, with rate $\lambda_A = 10$ calls/hour.

The system reaches a condition of statistical equilibrium in which the states with the highest probabilities are those corresponding to all lines occupied or only a few lines free. Under this condition, we can assume that the lines become free according to another Poisson process, with rate $\lambda_F = 30$ calls/hour, independently of the system state.

(a) Represent the system as a Markov Chain and draw its state diagram. It is suggested to characterize the state in terms of the number of class A users waiting in the queue, when the lines are all busy, otherwise in terms of the number of free lines.
(b) Find the state probabilities at the equilibrium.
(c) Find the blocking probabilities π_A and π_B for calls of the two classes.
(d) Find the maximum value of class A call arrival rate λ_{Amax} that guarantees the stability of the system.
(e) Find the average number of free lines $\bar{N}_F$ under the assumption that, for a sufficiently long time interval, the rate of high priority call arrivals is $\lambda_A = 0$.

Chapter 4
Engineering Circuit-Switched Networks

Abstract This chapter describes the main teletraffic engineering concepts for circuit-switched networks, or more generally for networks with static bandwidth allocation to traffic flows. The first part of the chapter deals with the simple case of a stand-alone system with only one type of customers. The classical Erlang $\mathscr{B}$ and $\mathscr{C}$ formulas are introduced and applied to a number of problems, including dimensioning the trunk group between two telephone central offices or the number of operators in a call center. The second part discusses the engineering issues that raise when considering the conditions of different classes of customers, or a more complex network topology. These models are not easily solvable without the aid of numerical analysis, but still some general concepts may be derived from the simple numerical examples presented here.

Keywords Circuit switching · $\mathscr{M}/\mathscr{M}$ queue · $\mathscr{M}/\mathscr{M}/m/0$ queue · Blocking probability · Erlang $\mathscr{B}$ formula · Public Switched Telephone Network (PSTN) · Trunk group · $\mathscr{M}/\mathscr{M}/m$ queue · Improving server utilization · Utilization of the last server · Infinite queue size · Erlang $\mathscr{C}$ formula · Waiting time · Call center dimensioning · Multi-service network · Bandwidth sharing · Complete partitioning · Partial sharing · Complete sharing · Trunk reservation · Fixed routing network

4.1 Introduction

The circuit switching paradigm has two main characteristics:

- static bandwidth allocation;
- signaling mostly taking place before and after the actual information exchange.

The typical and most classic example is the so-called Public Switched Telephone Network (PSTN), where phone calls are allocated a fixed bandwidth (64 Kbit/s, for instance), also called a *circuit*. The circuit is established during the call set-up phase, before the information exchange, and requires both user-to-network and network-to-network signaling. After the call set-up phase, if the circuit can be successfully

© Springer Nature Switzerland AG 2023
F. Callegati et al., *Traffic Engineering*, Textbooks in Telecommunication
Engineering, https://doi.org/10.1007/978-3-031-09589-4_4

established, a conversation phase follows, during which the network operations are fully transparent. The circuit is released once the conversation is over by means of a proper tear-down signaling phase.

Circuit switching systems are usually modeled as queuing systems with a finite number of servers, which represent the available circuits to be allocated to incoming calls. Circuits are typically assumed as full-duplex communication channels. Circuit switching is also characterized by the fact that, if all circuits are busy, the call request cannot be satisfied. Therefore it has to be either blocked or delayed, with a consequent service disruption experienced by the customer. Call set-up blocking or delay probabilities are the typical performance metrics of interest, as discussed in the following sections.

In this chapter we split the teletraffic study on circuit switching into three sections. At first, in Sect. 4.2 we present a stand-alone system without queuing space, the most typical circuit switching model. In particular, we study the $\mathcal{M}/\mathcal{M}$ and the $\mathcal{M}/\mathcal{M}/m/0$ queuing systems, deriving the famous Erlang $\mathcal{B}$ formula. Typical case studies of engineering and optimizing telephone central offices or trunk groups are presented. Then in Sect. 4.3 we consider stand-alone systems that are capable of queuing incoming calls when no servers are available. We study the $\mathcal{M}/\mathcal{M}/m$ system and its typical application to call center engineering. Finally, in Sect. 4.4 some more complex models are presented to consider multi-service traffic and circuit-switched networks with fixed routing, including related case studies.

These models provide a full coverage of the most typical problems that raise in the circuit switching scenario, obviously simplified to be accessible to the intended reader of this book.

4.2   Modeling Circuit Switching Systems Without Waiting Space

4.2.1   Performance Metrics

Before starting with the analysis, let us ask ourselves what are the relevant performance metrics we are targeting with such a study. Since the systems we are considering in this section do not have any queuing space, the waiting time is not an issue. By default the waiting time is 0 for every customer. Nonetheless, the system cannot accommodate all customer requests. Sometimes a customer arrives when all servers are busy and is not allowed to enter the system, given that all resources are taken. This is a congestion condition generally called a *blocking event*. The *blocking probability* is the major performance metric for this kind of systems. We call π_p the blocking probability and mostly focus on it as an engineering target.

However, if we focus on a public circuit switching system, like the traditional PSTN, there is another way to look at the network performance. The PSTN is operated by telecommunication infrastructure providers (typically called *Telco operators*) that deploy and maintain the network architecture, offering its use as a service to public customers at a price that is usually related to the customer traffic volume (see, for instance, the example in Sect. 2.6.1).

The blocking probability is the typical performance metric relevant to the customer, who would obviously like to experience the lowest possible π_p. But the performance metric of interest to the operator (besides the cost of the infrastructure, its resilience, the cost of maintenance, and other operational and financial quantities) is the resource utilization. When customers are charged based on service usage, a larger resource utilization will result in higher income, and possibly higher revenues. On the contrary, lower utilization means not only lower incomes and lower revenues, but could also lead to bankruptcy at some stage.

Therefore, the server utilization ρ is another important parameter to be considered here. A question that should be answered is whether these two quantities can be decoupled or are strictly bound to each other. Clearly, if they are correlated, a trade-off between user satisfaction and economic sustainability has to be found without a specific engineering effort, but if there is the possibility to decouple the two quantities, the network engineer should focus on optimizing both user satisfaction and operator revenues with proper engineering solutions. The latter issue will be discussed in the following, in particular in Sect. 4.2.3.2.

4.2.2 An Ideal System with Infinite Circuits

Now let us consider a system offering infinite service resources to the customers. It is indeed the most ideal queuing system we can think about. In this case incoming customers will always find a server free at once, without the need to wait or the risk to be blocked. It is rather obvious that this system is never congested.

If the user arrival process is a Poisson process and the service is time exponentially distributed (i.e., memoryless), the system is an $\mathcal{M}/\mathcal{M}$ queue and can be described as a BD process with the following rates:

$$
\begin{aligned}
\lambda_k &= \lambda \quad \forall k \geq 0 \\
\mu_k &= k\mu \quad \forall k > 0
\end{aligned}
\tag{4.1}
$$

The state probabilities (3.11) of the BD process in this case can be written as follows:

$$
P_k = P_0 \frac{\lambda^k}{k!\mu^k} = P_0 \frac{A_0^k}{k!} \quad \forall k > 0
\tag{4.2}
$$

with the normalizing constant:

$$P_0 = \left(\sum_{k=0}^{\infty} \frac{\lambda^k}{k! \mu^k} \right)^{-1} = \left(\sum_{k=0}^{\infty} \frac{A_0^k}{k!} \right)^{-1} = e^{-A_0} \tag{4.3}$$

Therefore, the full equation for the steady state probabilities is:

$$P_k = \frac{A_0^k}{k!} e^{-A_0} \quad \forall k \geq 0 \tag{4.4}$$

It is interesting to note that the steady state probabilities in this case take the form of the Poisson formula. This is to say that the number of customers in the system has a Poisson behavior, i.e., it is memoryless, which is rather intuitive since in a system like this what a given customer experiences is totally independent of other customers. Every customer gets access to a server upon arrival to the system, and the time spent in the system by a customer depends on the service time only, regardless of what happens to other customers.

4.2.2.1   Average Values

From the steady state probabilities it is easy to obtain the traffic in the system and the throughput. In this special case, they are the same quantity, considering that $\lambda_p = 0$ and $\lambda_s = \lambda$. Therefore:

$$A = \lambda \bar{\delta} = \lambda \bar{\vartheta} = A_0 = A_s \tag{4.5}$$

which may be also calculated as:

$$A = \sum_{k=0}^{\infty} k P_k = e^{-A_0} A_0 \sum_{k=1}^{\infty} \frac{A_0^{k-1}}{(k-1)!} = A_0 \tag{4.6}$$

and obviously:

$$\bar{\delta} = \frac{A}{\lambda} = \bar{\vartheta} \tag{4.7}$$

$$\bar{\eta} = 0 \tag{4.8}$$

4.2.2.2   What About Congestion?

As already outlined, in this case the system can always serve an incoming customer. Therefore, the notion of congestion does not have any real meaning. Nonetheless, it

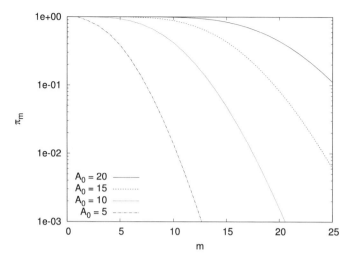

Fig. 4.1 $\mathscr{A}(m, A_0)$ as a function of m, varying A_0 as a parameter

may be useful to know the probability of having at least m customers in the systems, that is:

$$\pi_m = \sum_{k=m}^{\infty} \frac{A_0^k}{k!} e^{-A_0} = 1 - \sum_{k=0}^{m-1} \frac{A_0^k}{k!} e^{-A_0} = \mathscr{A}(m, A_0) \qquad (4.9)$$

Equation (4.9) is written as $\mathscr{A}(m, A_0)$ because it depends on m and A_0 only. Figure 4.1 shows its behavior as a function of m, varying A_0 as a parameter.

The reader may rightly be puzzled about what engineering use formula (4.9) can have. As a matter of fact, the ideal $\mathscr{M}/\mathscr{M}$ system is always able to satisfy the users expectations and no QoS degradation ever happens. However, this formula can be used as a performance benchmark for more realistic systems. For this reason it was used in the past as an approximate tool to dimension circuit-switched networks with some margin of additional safety. Equation (4.9) is sometimes also referred to as *Erlang $\mathscr{A}$* formula and gives the congestion probability of a queuing system where impatient customers abandon the queue with the same rate as the service completion, as will be discussed in Sect. 4.3.4.4.

4.2.3 The Real System with a Finite Number of Circuits

We now consider a more realistic circuit switching system, with m servers and without waiting space. This is called a *pure loss system*. It was extensively applied to design circuit-switched networks for the old analog telephony service. As usual, let us assume Poisson arrivals and exponential service times. This is an $\mathcal{M}/\mathcal{M}/m/0$ queuing system, and we can study it as a BD process with the following rates:

$$
\begin{aligned}
\lambda_k &= \lambda \quad 0 \le k < m \\
\mu_k &= k\mu \quad 0 < k \le m
\end{aligned}
\tag{4.10}
$$

Therefore:

$$
P_k = P_0 \frac{A_0^k}{k!} \quad 0 < k \le m
\tag{4.11}
$$

and:

$$
P_0 = \left(\sum_{k=0}^{m} \frac{A_0^k}{k!} \right)^{-1}
\tag{4.12}
$$

It is worth noting that the steady state probabilities of this system always exist, since formula (4.11) provides positive values for any A_0 and any m, such values are always less than 1, and they sum up to 1, given the way they are defined.

4.2.3.1 Congestion: The Erlang $\mathcal{B}$ Formula

According to the PASTA property we can write:

$$
\Pr\{\text{all servers busy} \mid \text{customer arrives}\} = \Pr\{\text{all servers busy}\} = \Pr\{k = m\}
$$

and then the blocking probability is:

$$
\pi_p = P_m = \frac{A_0^m / m!}{\sum_{k=0}^{m} A_0^k / k!} = \mathcal{B}(m, A_0)
\tag{4.13}
$$

Equation (4.13) is called the *Erlang $\mathcal{B}$* formula, written as $\mathcal{B}(m, A_0)$. Some values of $\mathcal{B}$ are plotted in Fig. 4.2 as a function of A_0, varying m as a parameter.

Obviously, π_p increases with A_0, as more requests lead to higher server occupancy and therefore to more frequent blocking events. What is interesting to note in the figure is the specific behavior of the curve. In particular, we can see that when $A_0 \ll m$ the blocking probability shows high variability as a function of A_0. This means that under these conditions the system performance is very sensitive to the offered traffic, and therefore a careful dimensioning of the system is necessary. When A_0 gets close to m (or even larger), the amount of blocked traffic is overwhelming and the performance is very poor, behaving almost independently of A_0. In this case the system is experiencing severe congestion and the quest for a specific grade of service is almost impossible to satisfy.

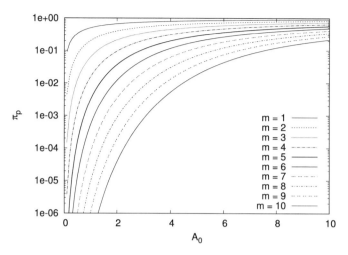

Fig. 4.2 $\mathscr{B}(m, A_0)$ as a function of A_0, varying m as a parameter

It can be proved that the Erlang $\mathscr{B}$ formula may be computed by means of the following recursive formula:

$$\mathscr{B}(m, A_0) = \frac{A_0 \, \mathscr{B}(m-1, A_0)}{m + A_0 \, \mathscr{B}(m-1, A_0)} \tag{4.14}$$

starting with $\mathscr{B}(0, A_0) = 1$. This is a very convenient method for automated computation.

4.2.3.2   Average Values and Utilization

Now let us find the traffic in a pure loss system. Considering that there is no queue, $A_c = 0$ and then:

$$A = A_s = (1 - \pi_p)A_0 = [1 - \mathscr{B}(m, A_0)] \, A_0 \tag{4.15}$$

where:

$$\lambda_s = (1 - \pi_p)\lambda \tag{4.16}$$

and:

$$\bar{\delta} = \frac{A}{\lambda_s} = \frac{(1 - \pi_p)A_0}{(1 - \pi_p)\lambda} = \frac{A_0}{\lambda} = \bar{\vartheta} \tag{4.17}$$

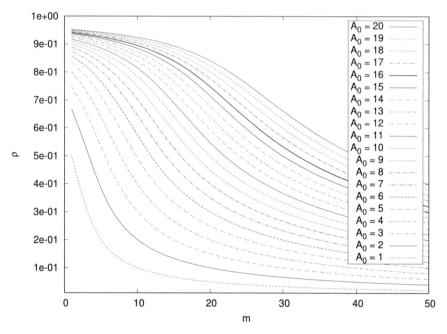

Fig. 4.3 ρ as a function of m, varying A_0 as a parameter

It follows that the server utilization is:

$$\rho = \frac{A_s}{m} = \frac{A_0(1 - \pi_p)}{m} \tag{4.18}$$

Figure 4.3 plots ρ as a function of m, varying A_0 as a parameter from 1 to 20. The figure is very interesting and provides a number of insights about the behavior of ρ. First of all, it says that the server utilization is higher for smaller values of m, once A_0 is fixed, and for larger values of A_0, once m is fixed. This is intuitively correct, as a higher offered traffic compared to a given number of servers means a higher arrival rate or a longer service time. In both cases, the servers will be asked to work more, either because of more frequent arrivals or because of longer service times. However, ρ tends to rapidly decrease to rather low values, well below 50%, when m increases. Such a behavior is relevant because, following the discussion in Sect. 4.2.1, the performance metrics of interest in this system are both π_p and ρ, providing the twofold view of customer and operator satisfaction level. It is also interesting to note that, for larger values of A_0 and a fixed ratio A_0/m, ρ is larger. This seems to suggest that larger systems with a larger number of servers tend to keep them busier. To better understand this last point, it is useful to observe two additional graphs that plot m and ρ as a function of A_0, showing curves where a given maximum value of the blocking probability π_p is guaranteed. In other words, Figs. 4.4 and 4.5 put together in the same graph the two performance metrics of

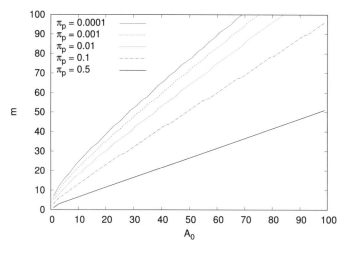

Fig. 4.4 Smallest value of m, such that the blocking probability is kept below π_p, as a function of A_0, varying π_p as a parameter

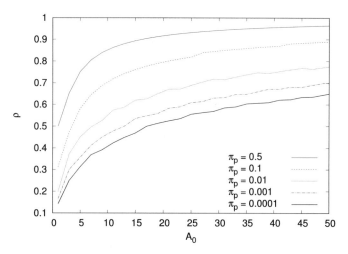

Fig. 4.5 ρ as a function of A_0 when m is assigned the smallest value such that the blocking probability is kept below π_p, varying π_p as a parameter

interest, and therefore can be helpful in suggesting the best engineering guidelines to guarantee both network customer and operator satisfaction.

In more details, in Fig. 4.4 the minimum number of servers that are required to guarantee a given blocking probability is plotted as a function of A_0, varying π_p as a parameter spanning different orders of magnitude. Obviously, m increases with A_0 but, since the number of servers is an integer, its variations occur with discrete units. This is the reason why the curves are not smooth but show a stepwise behavior, which occurs whenever the value of m must be increased by one unit to satisfy the given blocking probability value. From the figure it can be seen that, given a target π_p, the number of required servers does not increase proportionally to A_0. For instance, if we double A_0 from 20 E to 40 E for a target $\pi_p = 1\%$, the minimum number of required servers grows approximately from 30 to 50, i.e., it grows but it does not double.

The same message is conveyed in another form by Fig. 4.5, which is built in such a way that each curve representing ρ is plotted as a function of A_0 by choosing the smallest value of m that allows to keep the blocking probability below π_p, varying π_p as a parameter. Once A_0 and π_p are fixed, the throughput of the system is also fixed to $A_s = A_0(1 - \pi_p)$. Therefore, if the minimum number of required servers grows less than proportionally to A_0, it means that each server has to work more when A_0 increases, and therefore their individual utilization must increase, which is exactly what we observe in the graph.

So what is the lesson we learned? These results tell us that, because of the random fluctuation of the traffic, smaller systems with a low offered load must keep a number of servers idle (i.e., waiting for customers) that is higher in proportion to the offered traffic itself and, as a consequence, their utilization is inherently low. Larger systems receive more requests overall and may accommodate congestion periods with a number of servers that is lower in percentage with respect to A_0. As a consequence, their servers are better utilized while still providing the same level of customer satisfaction. This is a rather general rule of thumb in queuing systems: sharing a large number of resources among a large number of customers is generally better than sharing a few servers among a few customers. A practical example from the real life is the roundabout. We all know from practical experience that the waiting time at an intersection tends to be smaller when the intersection traffic is regulated with a roundabout compared to the case of traffic lights.

Nonetheless, performance improvements usually do not come for free. In systems with large A_0 the servers experience higher utilization, but this also means that there are less spare resources in case they are needed. If a system experiences an overload with an unexpected increase of the offered traffic, a system with large A_0 proves to be less resilient than a system with small A_0. This is shown in Fig. 4.6 where π_p is plotted as a function of the percentage increase of the offered traffic, considering three initial values of A_0. The graph shows that π_p increases more when the starting value of A_0 is larger, given that the percentage of increase is the same. This brings to the conclusion that systems with very large values of A_0 have to be designed more carefully, and should be slightly overdimensioned if a traffic overload is likely to happen.

4.2.4 Utilization of Ordered Servers

Up to now, following the basic assumptions presented in Sect. 1.3, we have imagined that the servers are all identical and the customers choose one of them at random when they arrive and find more than one server free. Now let us imagine a different behavior of the system and that the following three assumptions hold:

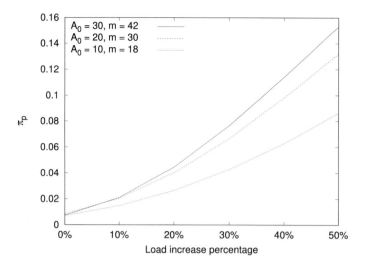

Fig. 4.6 π_p as a function of a percentage increase of A_0 with different starting values of A_0 and fixed values of m

1. the servers are numbered from 1 to m;
2. a customer is forced to choose the free server with the lowest index;
3. when a customer completes its service and leaves server number i, if there is a customer being served by server $i + 1$, it is transferred to server i, if there is a customer being served by server $i + 2$, it is transferred to server $i + 1$, and so on; in other words, if there are k customers in the system, they always keep the first k servers busy;
4. the transfer of a customer from any server $i + 1$ to server i happens instantly and does not change the residual service time of the customer.

This scheme may be easy to apply or not, depending on the real system implementation. Nonetheless, it is useful to study the utilization of the individual server. What we want to evaluate is the utilization of the i-th server, named ρ_i in the following, considering that the workload is not uniformly shared among the servers anymore. To obtain ρ_i let us recall that in $\mathcal{M}/\mathcal{M}/m/0$ systems there is a fraction of the offered traffic that is lost: $A_p = A_0 \mathcal{B}(m, A_0)$. Now, for a given i, let us imagine a system working with $m = i - 1$ ordered servers, as described above. The traffic lost by such system is $A_{p,i-1} = A_0 \mathcal{B}(i - 1, A_0)$. Similarly, if we consider a system with $m = i$, the traffic lost is $A_{p,i} = A_0 \mathcal{B}(i, A_0)$. Obviously $A_{p,i-1} > A_{p,i}$, since less servers carry less traffic. Now, given the ordered use of the severs, we can observe that the fraction of traffic carried by the system with $m = i$ exceeding the traffic carried by the system with $m = i - 1$ must be the traffic served by its i-th server. Therefore:

$$\rho_i = A_0 \left[\mathcal{B}(i - 1, A_0) - \mathcal{B}(i, A_0) \right] \tag{4.19}$$

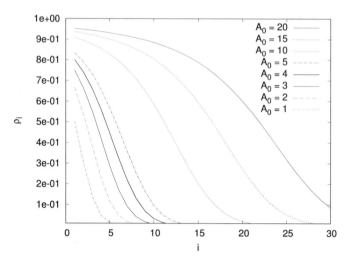

Fig. 4.7 Utilization ρ_i of the i-th ordered server as a function of i, varying A_0 as a parameter

The utilization of server i is plotted in Fig. 4.7 varying A_0 as a parameter. The graph confirms what discussed above. If we consider an ordered set of servers, we understand that a pure loss system like the $\mathcal{M}/\mathcal{M}/m/0$ has only one possible way to mitigate the blocking probability (i.e., the congestion of the system). Most of the time it has to keep some spare servers free, to be ready to serve the customers in those congestion periods that determine the blocking events. Therefore, if the target blocking probability is, for instance, $\pi_p \leq 1\%$, it means that the system needs a few servers that must be free for 99% of the time, therefore with very low utilization. These servers limit the average utilization of all servers, an effect that is more relevant in systems with a smaller number of servers, and which gets better if m increases. This leads once again to the conclusion that large systems carrying a lot of traffic make a better use of their resources.

4.2.5 Comparing the $\mathcal{M}/\mathcal{M}$ and the $\mathcal{M}/\mathcal{M}/m/0$ Systems

As already discussed above, the ideal system $\mathcal{M}/\mathcal{M}$ does not introduce any service degradation. Nonetheless, the Erlang $\mathscr{A}$ formula may be used as an estimate of service degradation in a real system of finite capacity m. Let us consider the same values for m and A_0 and compare the expressions of the Erlang $\mathscr{A}$ and $\mathscr{B}$ formulas. We can write:

$$\mathscr{A}(m, A_0) = \frac{\sum_{k=m}^{\infty} \frac{A_0^k}{k!}}{\sum_{k=0}^{\infty} \frac{A_0^k}{k!}} = \frac{\frac{A_0^m}{m!} + \sum_{k=m+1}^{\infty} \frac{A_0^k}{k!}}{\sum_{k=0}^{m} \frac{A_0^k}{k!} + \sum_{k=m+1}^{\infty} \frac{A_0^k}{k!}} \qquad (4.20)$$

Recalling that:

$$\mathscr{B}(m, A_0) = \frac{\frac{A_0^m}{m!}}{\sum_{k=0}^{m} \frac{A_0^k}{k!}} \qquad (4.21)$$

we can write:

$$\mathscr{B}(m, A_0) = \frac{x}{y} \qquad \mathscr{A}(m, A_0) = \frac{x+c}{y+c} \qquad (4.22)$$

where $c = \sum_{k=m+1}^{\infty} \frac{A_0^k}{k!}$. Comparing the expressions in (4.22), we see that the Erlang $\mathscr{A}$ formula is mathematically obtained from the Erlang $\mathscr{B}$ formula by summing the same quantity to both numerator and denominator. Given that $x < y$, we can conclude that:

$$\mathscr{B}(m, A_0) < \mathscr{A}(m, A_0) \qquad (4.23)$$

4.2.6   Insensitivity to Service Time Distribution

A very important property of the $\mathscr{B}(m, A_0)$ formula is its insensitivity to the service time distribution.

> If the service time is not exponentially distributed, the resulting $\mathscr{M}/\mathscr{G}/m/0$ system is not a Markov chain anymore, but the blocking probability can still be calculated using the Erlang $\mathscr{B}(m, A_0)$ formula.

4.2.7   How Good Is the Erlang $\mathscr{B}$ Model

As discussed in Sect. 2.2.1, assuming a Poisson arrival process implies the assumption of an infinite population of customers and full independence among their behaviors. It is not difficult to understand that in reality this cannot be true in general. In a real telephone network, the customer population is large but finite and correlations exist both inter and intra-customer. A typical example of inter-customer correlation is that, when a customer makes a call, there is another customer that is called, and that cannot start another call. Therefore, the behavior of the two customers is correlated. A typical example of intra-customer correlation is when a customer finds the callee's line busy. In this case a typical behavior is to retry after a short while for one or more times, until the callee is found available. This implies a correlation between the calls made by that specific customer.

The assumption of exponential service time distribution is also questionable. The exponential distribution implies that shorter service times are more likely. This is not true in general. Common experience says that, when we make a phone call, we need some time to say what we have to, and likely a call duration of a minute is more likely than a call of a few seconds. Similarly, we will never stay at the phone for a very long time, therefore the distribution of the call holding time will be truncated at some stage and very long calls will never occur.

These comments raise the question whether a $\mathcal{M}/\mathcal{M}$ model makes any sense when applied to real life. For what concerns the service time, the insensitivity of the Erlang $\mathcal{B}$ formula to the service time distribution solves the issue. To have an effective calculation of the relevant performance metrics, the service time distribution does not matter as long as we can correctly estimate its average. Concerning the arrival process, the answer is more complicated. The basic concept is simple but not easy to be tested in practice: the model makes sense if we can demonstrate that the above mentioned correlations as well as the finite population of customers have a negligible effect on the overall system behavior and performance with respect to the $\mathcal{M}/\mathcal{M}$ assumption. Intuitively speaking, it is easy to understand that a population that is finite but very large can be approximated as an infinite population. This is the case for large public telephone networks that account for very large numbers of subscribers, especially in the largely populated urban areas that represent the most challenging ones from the engineering and business point of view. Regarding the inter- and intra-customer correlation, intuition suggests that their effect might not be quantitatively very important if the correlated events happen more rarely compared to the overall number of events in the system. This is the case when we assume that each customer uses the system with a small probability, e.g., in the range of 1 to 10% at most. In this case, when considering a very large population, the correlated events will be only a few when compared to the uncorrelated events. This means that a customer calling for the first time finds the callee free most of the times, that a customer is rarely called exactly at the time when it starts a new call, etc. This is in line with the real life experience we all have of the telephone network.

4.2.8   Examples and Case Studies

4.2.8.1   Dimensioning the Number of Circuits in a PABX

A Private Automatic Branch Exchange (PABX) of a large company provides access to 100 users with an estimated offered traffic per user equal to $A_{0u} = 0.05$ E. The PABX does not provide queuing space for incoming calls and works according to the circuit switching principle. The connection between two users is established by means of a full-duplex circuit inside the central office. As long as all circuits are busy, any incoming call is blocked. Given that the target call blocking probability due to congestion must be $\pi_p \leq 1\%$, we wish to determine:

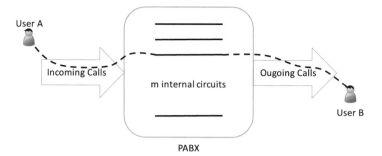

Fig. 4.8 The PABX as a loss system with m servers. The servers correspond to the internal PABX circuits that are used to interconnect customers when a call is established

1. the numbers of lines that are needed inside the switching exchange to guarantee the required blocking probability;
2. the utilization of the lines;
3. the blocking probability and utilization when one line becomes out of service;
4. the blocking probability when the offered traffic increases by 10%.

The PABX can be considered as a loss system, where the m servers correspond to the internal circuits used to interconnect customers when a call is established, as illustrated in Fig. 4.8. We assume that the overall traffic of calls requested by the users follows a Poisson process. The overall offered traffic is given by the sum of the offered traffic per user, i.e.:

$$A_0 = 0.05 \cdot 100 = 5 \text{ E} \tag{4.24}$$

We can use the Erlang $\mathscr{B}$ formula to obtain the required number of circuits such that:

$$\mathscr{B}(m, 5) \leq 0.01 \tag{4.25}$$

by iteratively applying the formula for increasing values of m:

$$[\ldots]$$

$$\mathscr{B}(8, 5) = 7.01 \cdot 10^{-2}$$

$$\mathscr{B}(9, 5) = 3.75 \cdot 10^{-2}$$

$$\mathscr{B}(10, 5) = 1.84 \cdot 10^{-2}$$

$$\mathscr{B}(11, 5) = 8.29 \cdot 10^{-3}$$

Therefore, the minimum number of circuits needed is:

$$m = 11 \tag{4.26}$$

With $m = 11$ the throughput is:

$$A_s = A_0 \left(1 - \pi_p\right) \cong 4.96 \text{ E} \tag{4.27}$$

and therefore the utilization of the circuits, in the assumption of random server selection, is:

$$\rho = \frac{A_s}{m} \cong 0.45 \tag{4.28}$$

As we already observed, for relatively small values of A_0 and m the average utilization of the servers is rather low, in the range of 50%.

Now if a circuit becomes out of service and cannot be used, we get:

$$\pi'_p = \mathscr{B}(10, 5) = 1.84 \cdot 10^{-2} \tag{4.29}$$

which is slightly above the threshold, but still in a rather acceptable range. The utilization grows a bit, but is still low:

$$\rho' = \frac{A_0(1 - \pi'_p)}{m} = 0.49 \tag{4.30}$$

Finally, if A_0 increases by 10%, with $m = 11$ we get:

$$\mathscr{B}(11, 5.5) = 1.44 \cdot 10^{-2} \tag{4.31}$$

i.e., the blocking probability increases significantly in relative terms (approximately +74%), but in absolute terms it is still in the range of the percent.

4.2.8.2   Planning the Dimensioning of a Trunk Group Between Two Central Offices

We are requested to plan the number of circuits in a "long distance trunk group" connecting two central offices, as illustrated in Fig. 4.9. The planning must be valid for a period of three years in the assumption that:

1. the average call arrival rate at the beginning of the first year is $\lambda(0) = 200$ calls/h, and at the end of the third year is expected to be $\lambda(3) = 1000$ calls/h;
2. during the three-year period, the call arrival rate increases exponentially from $\lambda(0)$ to $\lambda(3)$;
3. the average call holding time is $\bar{\vartheta} = 1.5$ min;

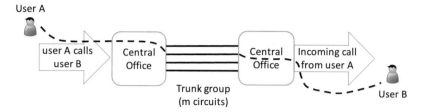

Fig. 4.9 The trunk group between two central offices to carry calls from users connected to one central office to users connected to the other

Table 4.1 Planning of a trunk group for a three-year period

Number of boards	Number of circuits	Max A_0 s.t. $\pi_p \leq 1\%$	λ (calls/h)	Time of system upgrade (years)	Utilization $\rho = A_s/m$
1	6	1.91	76		0.31
2	12	5.88	235	0.30	0.49
3	18	10.44	418	1.37	0.58
4	24	15.23	612	2.08	0.64
5	30	20.34	814	2.61	0.67
6	36	25.51	1020	3.00	0.71

4. the interconnection system between the two central offices is upgraded by adding boards that support $c = 6$ new circuits each;
5. the target call blocking probability is $\pi_p \leq 1\%$.

The exponential increase of the call arrival rate during the three-year period can be expressed as:

$$\lambda(t) = \lambda_0 \, e^{t/\tau} \quad 0 \leq t \leq 3 \tag{4.32}$$

Given that $\lambda(0) = \lambda_0 = 200$ calls/h and $\lambda(3) = 1000$ calls/h, it follows that:

$$\tau = \frac{3}{\ln\left(\lambda(3)/\lambda(0)\right)} = 1.86 \text{ years} \tag{4.33}$$

The complete planning is presented in Table 4.1, which reports the following quantities in the columns from first to last, respectively:

- number n_b of boards installed in each central office (the assumption is that the two central offices are equipped with the same circuit capacity);
- number of circuits $m = c \, n_b$ installed in each central office;
- maximum value of A_0 such that $\mathscr{B}(m, A_0) \leq 0.01$;
- average call arrival rate $\lambda = \frac{A_0}{\vartheta}$;
- time instant t_i when the system requires an upgrade from the number of boards currently installed to the number of boards specified in the next row, i.e., $t_i = \tau \ln\left(\frac{\lambda(t_i)}{\lambda_0}\right)$;
- average server utilization in the assumption of random server choice.

From the values in Table 4.1 we learn that the trunk should be installed at $t = 0$ with 2 boards (12 circuits). However, after only 4 months the system would require an upgrade to 3 boards (18 circuits) that will guarantee the required QoS for a rather long period, up to 1 year and 5 months. It may therefore be reasonable to install already 3 boards at $t = 0$. More in detail, we should compare the cost of the board today with the estimated cost of the board tomorrow plus the cost of the upgrade, and decide which of the two options is more convenient from the financial point of view. This sort of evaluations that include "management" issues are out of the scope of this book, but it is important to understand that teletraffic engineering provides only a partial answer to a careful and effective planning of the economics of the network operator.

Finally, it is worth noting, as already outlined, that with the increase of traffic and trunk size, the circuits utilization increases.

4.2.8.3   Dimensioning Interconnections in a Private Telephone Network

A company spread over four different sites, called A, B, C, and D, must install a private telephone network interconnecting the PABXs of the four sites. The idea is to permanently lease some lines from an operator such that the PABXs are connected through a full mesh network (i.e., each site is connected directly to each other site). Moreover, the PABXs must also be connected to the public network (PSTN) to guarantee the possibility to make and receive external calls. The company management asked the responsible technicians to perform a measurement campaign to evaluate the amount of traffic between the sites. The results of this campaign are reported in the following table, where the row and column indexes represent the source and destination nodes, respectively. This kind of representation is often called the *traffic matrix* of the system, and all values are expressed in Erlangs.

Source	Destination				
	A	B	C	D	PSTN
A		5	5	8	8
B	5		5	7	8
C	5	5		6	8
D	9	7	6		16
PSTN	7	7	7	15	

Let us call m_{XY} and A_{XY} the number of circuits and the offered traffic between sites X and Y, respectively. Also, let us call m_X and A_X the number of circuits and the offered traffic between site X and the public network, respectively. The goal of the project is to:

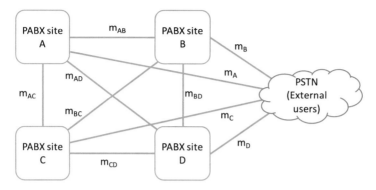

Fig. 4.10 Representation of the full mesh network with leased circuits among the four sites and external connections to the PSTN

1. dimension the full mesh network, i.e., m_{XY} and m_X for all sites, to guarantee a blocking probability $\pi_p \leq 1\%$;
2. consider the alternative of a star topology only for the private network, with a PABX working as transit node for calls between sites not directly connected; then dimension the network comparing the four alternatives of having A, B, C, or D as the center of the star topology;[1]
3. assume a single connection point with the public network at the center of the star topology and make that a transit node for all the inbound and outbound external traffic.

The full mesh network set-up is illustrated in Fig. 4.10. First of all, we must consider that the total amount of traffic between two sites, for instance, A and B, is given by the sum of the calls from A to B and the calls from B to A. Any voice call requires a bidirectional link, and therefore it does not matter from which node a call is originated, it will occupy a circuit in both directions. Therefore, to properly dimension the number of required circuits between A and B, the overall offered traffic to be considered is $A_{AB} = A_{A \to B} + A_{B \to A} = 5 + 5 = 10$ E. Then the number of circuits per link can be easily found with the Erlang $\mathscr{B}$ formula:

$$\mathscr{B}(m_{AB}, A_{AB}) = \mathscr{B}(m_{AB}, 10) \leq 0.01 \to m_{AB} = 18$$

$$\mathscr{B}(m_{AC}, A_{AC}) = \mathscr{B}(m_{AC}, 10) \leq 0.01 \to m_{AC} = 18$$

$$\mathscr{B}(m_{AD}, A_{AD}) = \mathscr{B}(m_{AD}, 17) \leq 0.01 \to m_{AD} = 27$$

$$\mathscr{B}(m_{BC}, A_{BC}) = \mathscr{B}(m_{BC}, 10) \leq 0.01 \to m_{BC} = 18$$

$$\mathscr{B}(m_{BD}, A_{BD}) = \mathscr{B}(m_{BD}, 14) \leq 0.01 \to m_{BD} = 23$$

$$\mathscr{B}(m_{CD}, A_{CD}) = \mathscr{B}(m_{CD}, 12) \leq 0.01 \to m_{CD} = 20$$

---

[1] Assume that the call blocking probability for a call traversing two links can be approximated with the blocking probability on each of the traversed link, thus neglecting any correlation effect while traversing two links.

Overall, in this case the company needs 124 leased lines to complete the internal network infrastructure.

Then let us consider the circuits needed for the PSTN connections:

$$\mathcal{B}(m_A, A_A) = \mathcal{B}(m_A, 15) \leq 0.01 \rightarrow m_A = 24$$

$$\mathcal{B}(m_B, A_B) = \mathcal{B}(m_B, 15) \leq 0.01 \rightarrow m_B = 24$$

$$\mathcal{B}(m_C, A_C) = \mathcal{B}(m_C, 15) \leq 0.01 \rightarrow m_C = 24$$

$$\mathcal{B}(m_D, A_D) = \mathcal{B}(m_D, 31) \leq 0.01 \rightarrow m_D = 43$$

Overall the company needs 115 interconnection circuits with the public network.

Now let us move to the star topology. In this case, on a given link we must consider not only the traffic between the two nodes at the edges of the link but also the transit traffic between the node that is not working as the center of the star and all the other nodes. For instance, if we assume A as the center of the star, the link AB will carry the offered traffic between A and B, plus the traffic between B and C, and the traffic between B and D. Therefore $A_{AB} = 10 + 10 + 14 = 34$ E.

If we choose A as the center of the star we obtain:

$$\mathcal{B}(m_{AB}, A_{AB}) = \mathcal{B}(m_{AB}, 10 + 10 + 14) \leq 0.01 \rightarrow m_{AB} = 46$$

$$\mathcal{B}(m_{AC}, A_{AC}) = \mathcal{B}(m_{AC}, 10 + 10 + 12) \leq 0.01 \rightarrow m_{AC} = 44$$

$$\mathcal{B}(m_{AD}, A_{AD}) = \mathcal{B}(m_{AD}, 17 + 14 + 12) \leq 0.01 \rightarrow m_{AD} = 56$$

for a total of 146 circuits.

If we choose B as the center of the star we obtain:

$$\mathcal{B}(m_{BA}, A_{BA}) = \mathcal{B}(m_{BA}, 10 + 10 + 17) \leq 0.01 \rightarrow m_{BA} = 49$$

$$\mathcal{B}(m_{BC}, A_{BC}) = \mathcal{B}(m_{BC}, 10 + 10 + 12) \leq 0.01 \rightarrow m_{BC} = 44$$

$$\mathcal{B}(m_{BD}, A_{BD}) = \mathcal{B}(m_{BD}, 14 + 17 + 12) \leq 0.01 \rightarrow m_{BD} = 56$$

for a total of 149 circuits.

If we choose C as the center of the star we obtain:

$$\mathcal{B}(m_{CA}, A_{CA}) = \mathcal{B}(m_{CA}, 10 + 10 + 17) \leq 0.01 \rightarrow m_{CA} = 49$$

$$\mathcal{B}(m_{CB}, A_{CB}) = \mathcal{B}(m_{CB}, 10 + 10 + 14) \leq 0.01 \rightarrow m_{CB} = 46$$

$$\mathcal{B}(m_{CD}, A_{CD}) = \mathcal{B}(m_{CD}, 12 + 17 + 14) \leq 0.01 \rightarrow m_{CD} = 56$$

for a total of 151 circuits.

If we choose D as the center of the star we obtain:

$$\mathscr{B}(m_{DA}, A_{DA}) = \mathscr{B}(m_{DA}, 17 + 10 + 10) \leq 0.01 \rightarrow m_{DA} = 49$$

$$\mathscr{B}(m_{DB}, A_{DB}) = \mathscr{B}(m_{DB}, 14 + 10 + 10) \leq 0.01 \rightarrow m_{DB} = 46$$

$$\mathscr{B}(m_{DC}, A_{DC}) = \mathscr{B}(m_{DC}, 12 + 10 + 10) \leq 0.01 \rightarrow m_{DC} = 44$$

for a total of 139 circuits.

It is rather obvious, from the results obtained, that the solution requiring less resources is the one with D chosen as the center of the star, with 139 required lines. This can be intuitively understood considering that the traffic between the nodes that are not the center of the star is doubled, since it is carried by two links. Therefore, the topology that has as a center the node exchanging the highest amount of traffic with the others is also the one that will carry the least total traffic and will require the smallest number of circuits.

Regarding the circuits connected to the PSTN, nothing changes and the solution is the same as before.

On the other hand, if the center of the star, node D according to the solution above, is also working as a gateway for all traffic exchanged with the PSTN, it happens that the links between D and the other nodes must carry also the external traffic, and therefore:

$$\mathscr{B}(m_{DA}, A_{DA}) = \mathscr{B}(m_{DA}, 17 + 10 + 10 + 15) = \mathscr{B}(m_{DA}, 52) = \leq 0.01 \rightarrow m_{DA} = 66$$

$$\mathscr{B}(m_{DB}, A_{DB}) = \mathscr{B}(m_{DB}, 14 + 10 + 10 + 15) = \mathscr{B}(m_{DB}, 49) = \leq 0.01 \rightarrow m_{DB} = 63$$

$$\mathscr{B}(m_{DC}, A_{DC}) = \mathscr{B}(m_{DC}, 12 + 10 + 10 + 15) = \mathscr{B}(m_{DC}, 47) = \leq 0.01 \rightarrow m_{DC} = 61$$

The circuits of the private network are now 190, whereas the circuits connecting node D to the PSTN are:

$$\mathscr{B}(m_D, A_D) = \mathscr{B}(m_D, 15 + 15 + 15 + 31) = \mathscr{B}(m_D, 76) = \leq 0.01 \rightarrow m_D = 92$$

All these solutions are viable and not very different in terms of total number of circuits. Nonetheless, they change the balance between the private internal circuits and the circuits interconnecting the private network to the PSTN. The difference in price of these alternatives could make one solution more convenient than the others.

4.2.8.4 Utilization of the Last Server and Network Cost Optimization

A company owns two sites located 50 Km apart. Each site is equipped with a telephone network with a dedicated PABX. The company plans to connect the two sites to implement a whole corporate network. The traffic between the two sites is estimated to be $A_0 = 12$ E, and the company management foresees that such value will likely be stable over the next few years.

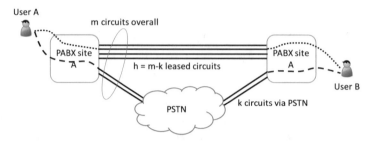

Fig. 4.11 A mixed trunk group between two PABXs: the m circuits of the trunk group are implemented by means of k lines switched through the PSTN and $h = m - k$ leased lines

The first problem is to dimension the number m of circuits to interconnect the two PABXs to guarantee a blocking probability per call $\pi_p \leq 0.01$. Moreover, the management is interested in knowing the average utilization ρ of such circuits.

The m circuits may be practically acquired in two ways, as illustrated in Fig. 4.11:

1. *leased lines* (type A), with a fixed cost per line per month C_l;
2. *switched lines* (type B), connected to the PSTN and used to call one site form the other; these lines imply a fixed service subscription cost[2] per month C_s, and an additional cost per Erlang of carried traffic C_0.

Given that m is the total number of circuits, let us call $h \leq m$ the number of circuits of type A and $k = m - h$ the number of circuits of type B. The goal is to dimension h and k in order to minimize the cost of the internal network. To do so, it is important to consider that the PABXs are intelligent and capable of keeping all the traffic on the leased lines as long as there are sufficient resources to carry it. This is to say that, for instance, if $h + 1$ circuits are in use and one of the calls carried by a type A line ends, the PABX automatically re-routes the call currently being carried by the type B line to the type A line that has become available. This functionality is called *call packing*.

The total number of required circuits m can be calculated with the Erlang $\mathcal{B}$ formula as usual:

$$\mathcal{B}(m, 12) \leq 0.01 \quad \Longrightarrow \quad m \geq 20 \tag{4.34}$$

and the server utilization is:

$$\rho = \frac{A_0(1 - \mathcal{B}(m, 12))}{m} = \frac{11.88}{20} = 0.594 \simeq 0,6 \tag{4.35}$$

---

[2]Let us assume that C_s is the cost per point of connection to the PSTN. If both PABXs need k connections to the PSTN, then C_s is charged $2k$ times.

In this case the monthly cost of a trunk group using only leased lines would be $C_A = m\,C_l$, whereas the monthly cost of a trunk group using only switched lines would be $C_B = m\,(2\,C_s + 0.6\,C_0)$. Therefore, the difference in cost between the two solutions can be obtained as

$$\Delta C = C_A - C_B = m\,(C_l - 2\,C_s - 0.6\,C_0) = m\,(C_f - 0.6\,C_0) \tag{4.36}$$

In this formula, $C_f = C_l - 2\,C_s$ is the net cost of a leased circuit when compared with a switched circuit: in practice, the fixed part of the cost of the switched circuit is subtracted from the fixed cost of the leased circuit. If $\Delta C < 0$ the leased circuits are more convenient, whereas if $\Delta C > 0$ the switched circuits are more convenient. C_f can be considered as the cost a user has to pay for having a circuit available 100% of the time, and C_0 the cost to pay to use a circuit only when needed, to be multiplied for the actual utilization. The former is a bulk-buy cost, for which we are used to the fact that when we buy more units we pay less per unit. Therefore, we expect that in general $C_f < C_0$, but how much? The difference between C_f and C_0 determines the economical trade-off as a function of the on-demand utilization of the circuit. In the specific case, if the bulk-buy discount is less than 40%, then the switched circuit is economically more convenient, otherwise the leased circuit is preferable.

Now let us consider the case of a system where the circuits are ordered and call packing can be used. We know already that, in this case, the server utilization is not uniform. Therefore, an optimization of the number of type A and type B circuits is required to minimize the overall cost. Let us imagine to number the circuits sequentially from 1 to m, and let us compute the traffic carried by each circuit when the call packing principle is applied. This is the situation considered in Sect. 4.2.4, therefore we can write:

$$\rho_i = A_0\,[\mathscr{B}(i-1, A_0) - \mathscr{B}(i, A_0)] \qquad i = 1, \ldots, m \tag{4.37}$$

The values of ρ_i for the specific case considered here are presented in Table 4.2. As expected, the circuits with small index i are used a lot (around 90% for $i \leq 4$), while the circuits with large index are used much less (in the range of 1% for the very last one). This is not surprising: indeed, according to the PASTA property, to guarantee a blocking probability $\pi_p \leq 1\%$ it is necessary that at least one server is free for almost 99% of the time.

The table also reports the portion of traffic carried by the h circuits of type A and the k circuits of type B under the assumption that $h = i$ and $k = m - i$, respectively:

$$A_{sA} = \sum_{j=1}^{h} \rho_j \qquad A_{sB} = \sum_{j=h+1}^{m} \rho_j \tag{4.38}$$

Obviously, there is also a portion of traffic that is blocked:

$$A_p = A_0\,\mathscr{B}(20, 12) = 12 \cdot 0.0098 = 0.118\ \text{E} \tag{4.39}$$

Table 4.2 The table shows the i-th server utilization for a system subject to $A_0 = 12$ E, with $m = 20$ ordered circuits and call packing, as well as the total traffic carried by the lines of type A and B under the assumption that $h = i$ and $k = m - i$

i	ρ_i	A_{sA} $(h = i)$	A_{sB} $(k = m - i)$
1	$9.2308 \cdot 10^{-1}$	0.9231	10.9594
2	$9.1222 \cdot 10^{-1}$	1.8353	10.0472
3	$8.9929 \cdot 10^{-1}$	2.7346	9.1479
4	$8.8385 \cdot 10^{-1}$	3.6184	8.2640
5	$8.6535 \cdot 10^{-1}$	4.4838	7.3987
6	$8.4315 \cdot 10^{-1}$	5.3269	6.5555
7	$8.1653 \cdot 10^{-1}$	6.1435	5.7390
8	$7.8467 \cdot 10^{-1}$	6.9281	4.9543
9	$7.4675 \cdot 10^{-1}$	7.6749	4.2076
10	$7.0201 \cdot 10^{-1}$	8.3769	3.5056
11	$6.4991 \cdot 10^{-1}$	9.0268	2.8556
12	$5.9038 \cdot 10^{-1}$	9.6172	2.2653
13	$5.2400 \cdot 10^{-1}$	10.1412	1.7413
14	$4.5229 \cdot 10^{-1}$	10.5935	1.2890
15	$3.7777 \cdot 10^{-1}$	10.9712	0.9112
16	$3.0380 \cdot 10^{-1}$	11.2750	0.6074
17	$2.3415 \cdot 10^{-1}$	11.5092	0.3733
18	$1.7228 \cdot 10^{-1}$	11.6815	0.2010
19	$1.2066 \cdot 10^{-1}$	11.8021	0.0803
20	$8.0303 \cdot 10^{-2}$	11.8825	0.0000

Now we can easily calculate the total cost of the network, for some generic values of h and k. The type A circuits will simply cost:

$$C_A = h\, C_l \tag{4.40}$$

whereas the cost of the switched lines will be:

$$C_B = 2\,k\, C_s + A_{sB}\, C_0 \tag{4.41}$$

Now let us assume to have a trunk group with h leased circuit, and then change the h-th circuit from leased to switched. We have to compare the case with h leased circuits and $k = m - h$ switched circuits with the case with $h' = h - 1$ leased circuits and $k' = m - h + 1$ switched circuits. The overall cost in the two cases is:

$$C_h = h\, C_l + 2\,(m - h)\, C_s + A_{sB}\, C_0$$
$$C_{h-1} = (h - 1)\, C_l + 2\,(m - h + 1)\, C_s + A'_{sB}\, C_0 \tag{4.42}$$

respectively, where:

$$A'_{sB} = \sum_{j=h}^{m} \rho_j \tag{4.43}$$

The cost difference is then given by:

$$\Delta C(h) = C_h - C_{h-1} = C_l - 2\,C_s + (A_{sB} - A'_{sB})\,C_0 \tag{4.44}$$

From (4.38) and (4.43) we can see that the difference in the traffic carried by type B circuits is given by the h-th server utilization:

$$A_{sB} - A'_{sB} = -\rho_h$$

Therefore, recalling that $C_f = C_l - 2\,C_s$:

$$\Delta C(h) = C_f - A_0\,[\mathscr{B}(h-1, A_0) - \mathscr{B}(h, A_0)]\,C_0 \tag{4.45}$$

$\Delta C(h)$ is a monotonic function of h and provides the solution to the optimization problem posed by the company management. Starting from a trunk group with $h = m = 20$ leased lines, if we replace the last leased circuit with a switched one, bringing h to $m - 1 = 19$, the cost difference is given by $\Delta C(m)$. If we then replace also the second to last leased circuit with a switched one, then the cost difference is $\Delta C(m - 1)$, and so on. Indeed, as long as $\Delta C(h) > 0$, the replacement of the h-th leased circuit with a switched one brings a cost saving. When $\Delta C(h) < 0$, the shift from leased to switched circuit is not worth anymore, because it would increase the overall cost of the infrastructure. Let away that, if $\Delta C(h) = 0$, the two options are equivalent in terms of cost. As an example, the values of $\Delta C(h)$ for a simple case with $A_0 = 12$ E and $C_0 = 1$ varying C_f from 0.2 to 0.6 are plotted in Fig. 4.12. The transition from negative to positive values corresponds to the value of h that minimizes the costs. The lower the value of C_f, the higher the optimal value of h, as intuition suggests. If the bulk-buy discount for a leased line is higher than the pay-per-use cost per Erlang, it is not convenient to have switched lines unless their utilization is really very low.

Finally, Fig. 4.13 shows the total cost of the trunk group under the same assumptions as in Fig. 4.12. As expected, the cost is a convex function of h with a minimum corresponding to the values when $\Delta C(h) = 0$. All curves start from the same value for $h = 0$ (no leased lines) because the traffic to be carried is the same for all cases. Then they open up and spread according to the value of C_f.

4.2.8.5   Coping with Traffic Increases

A trunk group between two PABXs includes $m = 80$ circuits that are sufficient to guarantee the required QoS of $\pi_p \leq 1\%$. Due to changes in the network architecture, the offered traffic in that network segment increases to a new value $A_0 = 80$ E. The problem is to determine the upgrade to the trunk group required to keep the QoS unchanged, which results in installing additional x circuits.

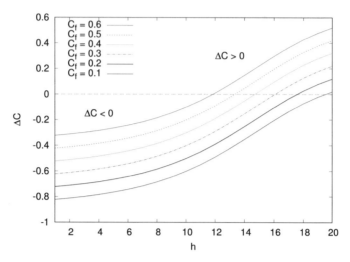

Fig. 4.12 Cost difference $\Delta C(h)$ when replacing the h-th leased line with a switched one as a function of h, with $A_0 = 12$ E, $C_0 = 1$ and $h + k = 20$, varying C_f as a parameter

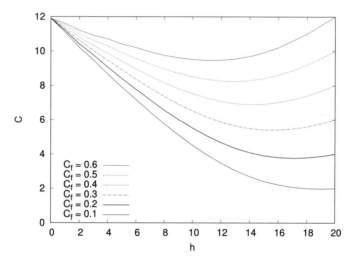

Fig. 4.13 Cost C of the trunk group interconnection as a function of h, with $A_0 = 12$ E, $C_0 = 1$ and $h + k = 20$, varying C_f as a parameter

To do that, we compare two possible operation alternatives:

1. the x circuits are simply added to the m existing, upgrading the hardware interconnecting the two central offices and the link in between;
2. the QoS objective is achieved by shifting the traffic blocked by the trunk group to an alternative path that is currently under implementation and that is equipped with the required additional x circuits; as a first step approximation, we assume that the traffic blocked from the trunk can be still considered distributed according to Poisson when offered to the alternative path.

If the assumptions are correct, we expect exactly the same value of x for the two alternatives, since the overall QoS objective and traffic carried by the network are the same.

At first let us find how many circuits are required in a single trunk group to carry $A_0 = 80$ E such that $\pi_p \leq 1\%$:

$$\mathscr{B}(m, 80) < 0.01 \quad \Longrightarrow \quad m = 80 + x \geq 96 \qquad (4.46)$$

Therefore:

$$x \geq 16$$

Now let us consider the second architectural alternative. The main trunk group with $m = 80$ circuits drops the following amount of traffic:

$$A_p = \mathscr{B}(80, 80)\, A_0 = 0.0841 \cdot 80 = 6.73 \text{ Erlang} \qquad (4.47)$$

If this blocked traffic is re-routed to the alternative path, under the simplifying assumption mentioned above the number of required circuits can be obtained as:

$$\mathscr{B}(x, 6.73) < 0.01 \quad \Longrightarrow \quad x \geq 14 \qquad (4.48)$$

The difference between the two results is somewhat surprising: why should the second alternative lead to a smaller value of x than the former, given that the traffic carried and QoS obtained are the same? The answer to this question must be sought in the solution adopted and, obviously, in the approximation made. The dimensioning of x in the former case is exact, since it is a simple application of the Erlang $\mathscr{B}$ formula. On the contrary, the latter solution is not exact because the blocked traffic offered to the additional circuits was assumed to be Poisson traffic.

As a matter of fact, it is known that the traffic "blocked" by a circuit switching system is not distributed according to Poisson, because it is affected by internal correlation. Calls are rejected because the system is congested and, therefore, calls arriving aside the rejected ones are more likely to be rejected as well, until there is a period with less arrivals and the congestion is solved. This is usually called a *peak* traffic period that tends to temporarily cause higher congestion with respect to the average Poisson load, because multiple arrivals happen closer to each other. Using the Erlang $\mathscr{B}$ formula in Eq. 4.48 means to underestimate the actual congestion level experienced by the alternative trunk group. Therefore, the solution with $x = 16$ is the correct one, and we should still use 16 circuits also in the alternative path to guarantee a correct QoS.

4.3 Modeling Circuit Switching Systems with Waiting Space

If we focus on a pure circuit switching system, like those that were once implemented in the traditional analog telephone network, queuing is a concept that does
not apply. Circuits were established by humans and/or automatic electro-mechanical
systems that acted at call set-up time, without the possibility to put a call on hold
when all the lines are busy. Nowadays, the circuit switching paradigm, even in the
telephone network, is basically implemented by means of computers that digitize
the voice and then transfer it as a sequence of bits. The circuit is represented by
the static logical relations between time slots and traffic flows, as those defined in
the TDM-PCM or SDH systems. More in general, any digital transmission of coded
information on a dedicated resource in the time, space, or frequency domain can be
considered as circuit switching.

In this context, the call set-up is a process that is managed by a computer and
that can be delayed to some extent, if all lines are currently busy, waiting for one to
become available soon. Moreover, there is a whole class of circuit switching systems
where putting a call on hold is somewhat normal: the call centers. We all know by
experience that, when calling the call center of the customer care service of a shop
or of an airline, we are usually asked to wait for the first operator to become free,
and we patiently stay at the phone, in queue, usually listening to some nice music
or pre-recorded message.

The material presented in this section intends to provide modeling tools for these
specific use cases, where the customers of a circuit switching system, instead of
being immediately blocked, can access a waiting space when all servers are found
busy.

4.3.1 Performance Metrics

Obviously, if a waiting space is available, call blocking in the sense we addressed
before may still exist, but it is not as much the core of the problem. In these systems
a customer may at first be asked to wait in the queue. Therefore, the probability π_r
that such event happens has to be evaluated. Then, if the queuing space is finite,
some customers may be blocked because all servers are busy and no waiting place
in the queue is available. This case is usually less important, because systems of this
kind are often equipped with a rather large waiting space that makes the blocking a
very rare event.

For this reason, and for the sake of a simpler modeling approach, the assumption
that is usually made and that we also follow here is that the waiting space is
infinite. Therefore, $\pi_p = 0$ and the performance metrics of interest are related to the
queuing event. However, not only the probability π_r of such event is relevant. Other
performance metrics must be considered, such as the *waiting time*, also called *latency*.
For instance, we are interested in knowing the average waiting time, or obtaining the

waiting time probability distribution, etc. Moreover, as we will see in more detail in the following, the waiting time is somewhat a fuzzy parameter that has to be better qualified. In this book we consider two different definitions of waiting time: the *global waiting time*, called η, that is the waiting time considering all customers, and the *conditioned waiting time*, called ϵ, that is the waiting time considering only the customers that actually experience latency, leaving out the customers that are lucky enough to find a free server when they arrive at the system and do not have to wait in the queue.

The two quantities η and ϵ are obviously closely related, but not equal. Generally speaking, their difference is related to π_r: if the probability to wait is large, then η and ϵ tend to be very similar, whereas if the probability to wait is small, η and ϵ tend to be very different. Which of the two latency metrics is more significant for engineering purposes is a matter of discussion. Indeed, when we think of human customers, the probability to wait for a long time is disappointing and such disappointment in general is not compensated by the fact that many other customers are lucky and do not wait at all. This is to say that probably ϵ better describes the perception of service degradation by human customers. If customers are machines, then the story can be different and the average global time necessary to complete a task may be a more effective performance metric. In brief, whether η or ϵ is more important depends on the specific application of the queuing system, and therefore we need to provide tools to calculate both metrics. The remainder of this section provides quantitative tools to address these issues.

4.3.2   The $\mathcal{M}/\mathcal{M}/m$ System

The system to be considered in this case is similar to the previous one, but has an infinite state space. Therefore, the birth and death rates are:

$$\lambda_k = \lambda \quad \forall k \geq 0$$

$$\mu_k = \begin{cases} k\mu & 0 < k \leq m \\ m\mu & k > m \end{cases} \tag{4.49}$$

and the steady state probabilities take the form:

$$P_k = \begin{cases} P_0 \dfrac{A_0^k}{k!} & 0 \leq k < m \\ P_0 \dfrac{A_0^k}{m!\,m^{k-m}} & k \geq m \end{cases} \tag{4.50}$$

with:

$$P_0 = \left[\sum_{k=0}^{m-1} \frac{A_0^k}{k!} + \frac{A_0^m}{m!} \sum_{k=m}^{\infty} \left(\frac{A_0}{m}\right)^{k-m} \right]^{-1} \tag{4.51}$$

Similarly to what discussed in the example of Sect. 3.5.3, in this case we need to be careful with the calculation of P_0, because of the sum of infinite terms appearing in Eq. (4.51). It is well known that the sum of the geometric series $\sum_{k=0}^{\infty} x^k$ converges to $\frac{1}{1-x}$ only if $-1 < x < 1$. In the case of an $\mathcal{M}/\mathcal{M}/m$ system, this means $A_0/m < 1$. Considering that in a system with an infinite queue there is no blocked traffic, i.e., $A_p = 0$ and then $A_s = A_0$, the convergence condition for Eq. (4.51) can be expressed in terms of server utilization as follows:

$$\rho = \frac{A_0}{m} = \frac{\lambda}{m\mu} < 1 \tag{4.52}$$

This is a critical constraint that stems from mathematical equations, but that also provides some practical insight on the characteristics of the $\mathcal{M}/\mathcal{M}/m$ system. The maximum throughput in this case is m, when every server is working full-time without pause or vacation. Having an offered traffic $A_0 > m$ means that the system is requested to serve more traffic than what it is able to carry, and therefore the excessive traffic piles up in a queue that grows to infinity. As a consequence, the system does not reach a statistical equilibrium, it is neither stationary nor ergodic, and therefore the steady state probabilities are undefined. Mathematically speaking, this cause the geometric series in Eq. (4.51) to go to infinity, meaning that $P_0 = 0$ as well as $P_k = 0$ $\forall k$. This behavior is rather easy to understand.

However, the geometric series does not converge either when $A_0 = m$, which at a first sight could be considered logically acceptable: the system is under stress with servers requested to work full-time, but in principle all the offered traffic could be successfully served. While this reasoning is true for a deterministic system with a constant incoming load, it cannot be applied to a stochastic system. Because of the randomness of the Poisson arrival process, it could happen at some stage that less than the expected average arrival rate occurs for some time period. This could lead to one or more servers to temporarily stop working because of lack of customers, meaning that for some time the throughput of the system goes below m. From that moment on, the system works under unbalanced conditions. At some point the traffic increases and more customers than the average arrival rate request a service, to balance the reduced number of arrivals in the previous period. However, the system cannot increase its service capacity accordingly, which is bound to m as a maximum. As a consequence, the queue starts growing and growing and never depletes again, leading to a system without statistical stability.

From now on, we will always consider systems where the condition $A_0 < m$ holds, and for which we can calculate the steady state probabilities and the related performance metrics. Therefore, under this assumption we can write:

$$P_0 = \left(\sum_{k=0}^{m-1} \frac{A_0^k}{k!} + \frac{A_0^m}{m!} \frac{m}{m - A_0} \right)^{-1} \tag{4.53}$$

4.3.2.1 Congestion: The Erlang $\mathscr{C}$ Formula

Thanks to the PASTA property, we can obtain the probability of a customer being queued as:

$$\Pr\{\text{all servers are busy} \mid \text{customer arrives}\} = \Pr\{\text{all servers are busy}\} = \Pr\{k \geq m\}$$

and therefore:

$$\pi_r = \Pr\{k \geq m\} = \sum_{k=m}^{\infty} P_k = P_0 \frac{A_0^m}{m!} \sum_{k=m}^{+\infty} \left(\frac{A_0}{m}\right)^{k-m}$$

$$= \frac{\frac{A_0^m}{m!} \frac{m}{m-A_0}}{\sum_{k=0}^{m-1} \frac{A_0^k}{k!} + \frac{A_0^m}{m!} \frac{m}{m-A_0}} = \mathscr{C}(m, A_0) \tag{4.54}$$

Formula (4.54) is called the Erlang $\mathscr{C}$ formula. Similarly to the Erlang $\mathscr{B}$ formula, the $\mathscr{C}$ formula can be calculated numerically. The easiest way is to reuse the values of the Erlang $\mathscr{B}$, since we can show that:

$$\mathscr{C}(m, A_0) = \frac{\frac{A_0^m}{m!} \frac{m}{m-A_0}}{\sum_{k=0}^{m-1} \frac{A_0^k}{k!} + \frac{A_0^m}{m!} \frac{m}{m-A_0}} = \frac{\frac{A_0^m}{m!} \frac{m}{m-A_0}}{\sum_{k=0}^{m} \frac{A_0^k}{k!} + \frac{A_0^m}{m!} \frac{A_0}{m-A_0}}$$

$$= \frac{\frac{m}{m-A_0}}{\frac{1}{\mathscr{B}(m,A_0)} + \frac{A_0}{m-A_0}} = \frac{m\mathscr{B}(m, A_0)}{m - A_0 + A_0\mathscr{B}(m, A_0)} \tag{4.55}$$

Solving equation (4.14) for A_0 we obtain:

$$A_0 = \frac{m\mathscr{B}(m, A_0)}{\mathscr{B}(m - 1, A_0)(1 - \mathscr{B}(m, A_0))} \tag{4.56}$$

and therefore:

$$\mathscr{C}(m, A_0) = \frac{m\mathscr{B}(m, A_0)}{m - A_0 + A_0\mathscr{B}(m, A_0)} = \frac{1}{\frac{1}{\mathscr{B}(m,A_0)} - \frac{1}{\mathscr{B}(m-1,A_0)}} \tag{4.57}$$

To easily obtain numerical values of the Erlang $\mathscr{C}$ formula, the latter expression can be included in any program or spreadsheet used to recursively calculate the Erlang $\mathscr{B}$.

4.3.2.2 Average Number of Customers

The throughput of the system can be easily obtained considering that no customer is lost. Therefore:

$$\lambda_s = \lambda$$

and:

$$A_s = \lambda \bar{\vartheta} = A_0 \tag{4.58}$$

The average number of customers waiting in the queue can be found using the steady state probabilities, considering that when k customers are in the system, with $k \geq m$, then $h = k - m$ customers are waiting in the queue. Therefore:[3]

$$A_c = \sum_{k=m}^{\infty} (k - m) P_k = \frac{P_0 A_0^m}{m!} \sum_{k=m}^{\infty} (k - m) \frac{A_0^{k-m}}{m^{k-m}} = \frac{P_0 A_0^m}{m!} \sum_{h=0}^{\infty} h \rho^h \tag{4.60}$$

and finally:

$$A_c = P_0 \frac{A_0^m}{m!} \frac{\rho}{(1-\rho)^2} = P_0 \frac{A_0^m}{m!} \frac{m}{m - A_0} \frac{A_0}{m - A_0} = \mathscr{C}(m, A_0) \frac{A_0}{m - A_0} \tag{4.61}$$

Knowing A_s and A_c makes it easy to calculate the traffic in the system:

$$A = \sum_{k=0}^{\infty} k P_k = A_s + A_c = A_0 + \frac{A_0}{m - A_0} \mathscr{C}(m, A_0) \tag{4.62}$$

and the average time spent in the system:

$$\bar{\delta} = \frac{A}{\lambda} = \bar{\vartheta} + \frac{\bar{\vartheta}}{m - A_0} \mathscr{C}(m, A_0) = \bar{\vartheta} + \bar{\eta} \tag{4.63}$$

---

[3] The series $\sum_{k=0}^{\infty} k x^k$ converges for $-1 < x < 1$. It can be shown as follows:

$$\sum_{k=0}^{\infty} k x^k = x \sum_{k=0}^{\infty} k x^{k-1} = x \sum_{k=0}^{\infty} \frac{d}{dx} x^k = x \frac{d}{dx} \sum_{k=0}^{\infty} x^k = x \frac{d}{dx} \frac{1}{1-x} = \frac{x}{(1-x)^2} \tag{4.59}$$

4.3.2.3 Average Waiting Time

We can now calculate the average waiting time considering all customers $\bar{\eta}$ (i.e., the global waiting time as defined above), as well as the average waiting time conditioned to being queued $\bar{\varepsilon}$. To obtain the former we can use Little's Theorem:

$$\bar{\eta} = \frac{A_c}{\lambda} = \frac{\mathscr{C}(m, A_0)}{\lambda} \frac{A_0}{m - A_0} = \mathscr{C}(m, A_0) \frac{\bar{\vartheta}}{m - A_0} \tag{4.64}$$

To obtain the latter we must consider the state probabilities $P_{k/k \geq m}$ conditioned to the fact that all servers are busy, and then any arrival is sent to the queue. The average number of customers in the queue, considering only the congested states, is then given by:

$$A_{c|k \geq m} = \sum_{k=m}^{\infty} (k - m) P_{k/k \geq m} \tag{4.65}$$

Since when $k \geq m$ we can write $\Pr\{k, k \geq m\} = \Pr\{k\} = P_k$, recalling the properties of a conditional probability we have:

$$P_{k|k \geq m} = \frac{\Pr\{k, k \geq m\}}{\Pr\{k \geq m\}} = \frac{P_k}{\Pr\{k \geq m\}} = \frac{P_k}{\mathscr{C}(m, A_0)} \quad k \geq m \tag{4.66}$$

Finally:

$$A_{c|k \geq m} = \sum_{k=m}^{\infty} (k - m) P_{k|k \geq m} = \frac{1}{\mathscr{C}(m, A_0)} \sum_{k=m}^{\infty} (k - m) P_k = \frac{A_0}{m - A_0} \tag{4.67}$$

and therefore:

$$\bar{\varepsilon} = \frac{A_{c|k \geq m}}{\lambda} = \frac{\bar{\eta}}{\mathscr{C}(m, A_0)} = \frac{\bar{\vartheta}}{m - A_0} \tag{4.68}$$

It is important to note that so far we have not made any specific assumption about the scheduling policy, so all the results presented above *are valid for any scheduling policy*.

From Eqs. (4.64) and (4.68) we can derive some simple formulas to find the minimum number of servers m needed to meet some specific requirement on the waiting time. Let us consider the two different definitions of waiting time.

- If we have a constraint on the average global waiting time such as $\bar{\eta} \leq \eta_0$, where η_0 is a target global queuing time, then we must ensure that:

$$m \geq A_0 + \mathscr{C}(m, A_0) \frac{\bar{\vartheta}}{\eta_0} \tag{4.69}$$

- If we have a constraint on the average waiting time for customers that have to wait such as $\bar{\epsilon} \leq \epsilon_0$, where ϵ_0 is a target conditioned queuing time, then we must ensure that:

$$m \geq A_0 + \frac{\bar{\vartheta}}{\epsilon_0} \tag{4.70}$$

In both cases we see that, as expected, m must be larger than A_0, and that the required additional number of servers increases when the target queuing time decreases or the service time increases.

4.3.3   Waiting Time Probability Distribution for a FIFO Queue

Assuming a FIFO queuing discipline, it is possible to calculate the probability density functions of η, ε and δ.

Let us start from $f_{\eta/k}(t)$, i.e., the probability density function of η conditioned to the fact that a customer finds k other customers in the system when it arrives. From $f_{\eta/k}(t)$ it is possible to obtain the probability density function of η as follows:

$$f_\eta(t) = \sum_{k=0}^{\infty} f_{\eta/k}(t) P_k \quad t \geq 0 \tag{4.71}$$

If $0 \leq k < m$, $\eta = 0$ and we can write:

$$f_{\eta/k}(t) = \delta(t) \quad t \geq 0 \tag{4.72}$$

where $\delta(t)$ is Dirac delta distribution, defined as:

$$\delta(t) = \begin{cases} 0 & \forall t \neq 0 \\ \infty & t = 0 \end{cases} \tag{4.73}$$

If $k \geq m$, we can write $k = m + j$ with $j \geq 0$, and then:

$$f_{\eta/m+j}(t) = \frac{(m\mu)^{j+1}}{j!} e^{-m\mu t} t^j \quad t > 0 \tag{4.74}$$

The probability density function in (4.74) can be calculated recursively:

- When $j = 0$, all servers are busy and the queue is empty. A new customer is queued and has to wait that one of the m customers currently being served completes its service. This happens with rate $m\mu$ and, due to the memoryless property of the exponential service time, according to an exponential distribution, i.e.:

$$f_{\eta/m}(t) = m\mu e^{-m\mu t} \quad t > 0 \tag{4.75}$$

- When $j = 1$, all servers are busy and there is already one customer waiting in the queue. A new customer is queued and then must wait for two services to complete, one to allow the first customer in line to start the service, another to finally start its own service. If η_1 and η_2 are the times needed to complete the two services, respectively, then the waiting time for the newly arrived customer is $\eta = \eta_1 + \eta_2$. Due to the memoryless property of the exponential distribution, η_1 and η_2 are two independent random variables identically distributed according to (4.75). Therefore, the probability density function of η is given by the convolution of two exponential probability density functions with parameter $m\mu$, i.e.:

$$f_{\eta/m+1}(t) = m\mu e^{-m\mu t} \star m\mu e^{-m\mu t}$$

$$= \int_0^t m\mu e^{-m\mu \tau} \cdot m\mu e^{-m\mu(t-\tau)} d\tau = (m\mu)^2 e^{-m\mu t} t \quad t > 0 \tag{4.76}$$

- Similarly, when $j = 2$, it is necessary that 3 servers get free before a new customer can start its service. Therefore, the probability density function of η is given by the convolution of (4.76) and (4.75), i.e.:

$$f_{\eta/m+2}(t) = (m\mu)^2 e^{-m\mu t} t \star m\mu e^{-m\mu t}$$

$$= \int_0^t (m\mu)^2 e^{-m\mu \tau} \tau \cdot m\mu e^{-m\mu(t-\tau)} d\tau = (m\mu)^3 e^{-m\mu t} \frac{t^2}{2} \quad t > 0$$

$$\tag{4.77}$$

- In general, when j customers are already waiting in the queue, the waiting time of a new customer has the following probability density function:[4]

$$f_{\eta/m+j}(t) = f_{\eta/m+j-1} \star m\mu e^{-m\mu t}$$

$$= \int_0^t \frac{(m\mu)^j}{(j-1)!} e^{-m\mu \tau} \tau^{j-1} \cdot m\mu e^{-m\mu(t-\tau)} d\tau$$

$$= \frac{(m\mu)^{j+1}}{j!} e^{-m\mu t} t^j \quad t > 0 \tag{4.78}$$

---

[4]The expression of the probability density function (4.78) coincides with that of the Erlang distribution of grade $r = j + 1$, as defined in (2.40), with average $(j + 1)/(m\mu)$. This is clearly due to the fact that the waiting time of the customer that finds all servers busy and j customers already in the queue is a random variable resulting from the sum of $j + 1$ identical exponentially distributed random variables with parameter $m\mu$.

Summing for all values of k we obtain:

$$
\begin{aligned}
f_\eta(t) &= \sum_{k=0}^{m-1} f_{\eta/k}(t)\, P_k + \sum_{k=m}^{\infty} f_{\eta/k}(t)\, P_k \\
&= \delta(t) \sum_{k=0}^{m-1} P_k + \sum_{j=0}^{\infty} \frac{(m\mu)^{j+1}}{j!}\, e^{-m\mu t}\, t^j\, P_0 \frac{A_0^{m+j}}{m!\, m^j} \\
&= \delta(t)\, [1 - \mathscr{C}(m, A_0)] + P_0 m\mu \frac{A_0^m}{m!}\, e^{-m\mu t} \sum_{j=0}^{\infty} \frac{(A_0 \mu t)^j}{j!} \\
&= \delta(t)\, [1 - \mathscr{C}(m, A_0)] + P_0 m\mu \frac{A_0^m}{m!}\, e^{-m\mu t}\, e^{\lambda t} \\
&= \delta(t)\, [1 - \mathscr{C}(m, A_0)] + \mathscr{C}(m, A_0)\, (m\mu - \lambda)\, e^{-(m\mu - \lambda)t}
\end{aligned}
\tag{4.79}
$$

The corresponding probability distribution is:

$$
\begin{aligned}
F_\eta(t) &= [1 - \mathscr{C}(m, A_0)] \int_0^t \delta(\tau)\, d\tau + \mathscr{C}(m, A_0)\, (m\mu - \lambda) \int_0^t e^{-(m\mu - \lambda)\tau}\, d\tau \\
&= 1 - \mathscr{C}(m, A_0)\, e^{-(m\mu - \lambda)t}
\end{aligned}
\tag{4.80}
$$

In a similar way it is possible to calculate $f_\varepsilon(t)$ using $P_{k/k \geq m}$ instead of P_k and summing only for $k > m$, obtaining:

$$
f_\varepsilon(t) = \sum_{k=m}^{\infty} f_{\eta/k}(t)\, P_{k/k \geq m} = \sum_{k=m}^{\infty} f_{\eta/k}(t) \frac{P_k}{\mathscr{C}(m, A_0)} = (m\mu - \lambda) e^{-(m\mu - \lambda)t}
\tag{4.81}
$$

and then:

$$
F_\varepsilon(t) = 1 - e^{-(m\mu - \lambda)t}
\tag{4.82}
$$

At last, recalling that the Dirac delta distribution is such that:

$$
\delta(t) \star f(t) = \int_0^t \delta(\tau) f(t - \tau)\, d\tau = f(t)
$$

and that the total time spent in the system is $\delta = \vartheta + \eta$, we get:

$$f_\delta(t) = f_\vartheta(t) \star f_\eta(t)$$

$$= \int_0^t \left[(1 - \mathscr{C}(m, A_0))\delta(\tau) + \mathscr{C}(m, A_0)(m\mu - \lambda)e^{-(m\mu-\lambda)\tau}\right] \cdot \mu e^{-\mu(t-\tau)} d\tau$$

$$= [1 - \mathscr{C}(m, A_0)]\mu e^{-\mu t} + \mathscr{C}(m, A_0)(m\mu - \lambda)\mu e^{-\mu t} \int_0^t e^{-(m\mu-\lambda)\tau} e^{\mu\tau} d\tau$$

$$= [1 - \mathscr{C}(m, A_0)]\mu e^{-\mu t} + \mathscr{C}(m, A_0)\frac{(m\mu - \lambda)\mu e^{-\mu t}}{\lambda - (m-1)\mu}\left(e^{-(m\mu-\lambda-\mu)t} - 1\right)$$

$$= \left[1 - \frac{\mu\mathscr{C}(m, A_0)}{\lambda - (m-1)\mu}\right]\mu e^{-\mu t} + \frac{\mu\mathscr{C}(m, A_0)}{\lambda - (m-1)\mu}(m\mu - \lambda)e^{-(m\mu-\lambda)t} \qquad (4.83)$$

From the probability density functions found above it is possible to obtain the average values that are obviously the same already found by means of Little's Theorem.

Furthermore, from the probability distributions in Eqs. (4.80) and (4.82) we can derive some simple formulas to find the minimum number of servers m needed to meet some specific requirement on the waiting time statistics. In this case, the QoS requirement can be expressed in terms of the probability that the waiting time stays below a given target value. Considering the two different definitions of waiting time, we can obtain what follows:

- If we have a constraint on the distribution of the global waiting time such as $\Pr\{\eta \leq \eta_0\} \geq \pi_0$, where we call η_0 the limit queuing time and π_0 the target probability that the requirement is met, then we can write:

$$F_\eta(\eta_0) \geq \pi_0 \qquad (4.84)$$

and solving (4.80) for m we obtain:

$$m \geq A_0 + \frac{\bar\vartheta}{\eta_0} \ln\left(\frac{\mathscr{C}(m, A_0)}{1 - \pi_0}\right) \qquad (4.85)$$

- If we have a constraint on the distribution of the waiting time for the customers that have to wait such as $\Pr\{\epsilon \leq \epsilon_0\} \geq \pi_0$, where we call ϵ_0 the limit queuing time and π_0 the target probability that the requirement is met, then we can write:

$$F_\epsilon(\epsilon_0) \geq \pi_0 \qquad (4.86)$$

and solving (4.82) for m we obtain:

$$m \geq A_0 + \frac{\bar\vartheta}{\epsilon_0} \ln\left(\frac{1}{1 - \pi_0}\right) \qquad (4.87)$$

It is interesting to note the similarity of Eqs. (4.85) and (4.87) with Eqs. (4.69) and (4.70), targeting, respectively, the average waiting time and its distribution.

4.3.4 Examples and Case Studies

4.3.4.1 Dimensioning the Number of Operators in a Call Center

A University provides live information to the students by means of a call center. The University has 12 schools: 6 "large" ones with approximately 10,000 students and 6 "medium" ones with approximately 6000 students. The call center operates 5 hours per day, 25 days per month. Let us assume that the calls arrive according to a Poisson process with constant arrival rate over the hours of operations, and that the calls have random holding times that are independent and exponentially distributed with an average $\bar{\vartheta} = 3$ min. Furthermore, let us assume that on average each student in a month calls with probability 0.5.

We need to compare the number of operators that would be needed in the call center under the following possible alternatives:

1. deploy a call center for each school, working without queuing, according to a pure circuit switching principle, with the goal to guarantee a call blocking probability $\pi_p \leq 5\%$;
2. deploy a call center for each school, supporting queuing with FIFO scheduling when all operators are busy, with the goal to guarantee a waiting time for the customers that have to wait $\epsilon_0 \leq 3$ min with probability $\pi_0 \geq 95\%$;
3. deploy a single centralized call center for the whole University, supporting queuing, with the goal to guarantee the same quality of service as the case above.

The offered traffic per school can be obtained as follows:

- for "medium" schools the call arrival rate is:

$$\lambda_m = \frac{6000 \cdot 0.5}{25 \cdot 5} = 24 \text{ calls/h} = 0.4 \text{ calls/min}$$

and, since the average call holding time is $\bar{\vartheta} = 3$ min, we have:

$$A_{0m} = \lambda_m \cdot \bar{\vartheta} = 1.2 \text{ E}$$

- for "large" schools the call arrival rate is:

$$\lambda_l = \frac{10000 \cdot 0.5}{25 \cdot 5} = 40 \text{ calls/h} = 0.667 \text{ calls/min}$$

and then:

$$A_{0l} = \lambda_l \cdot \bar{\vartheta} = 2 \, \text{E}$$

Now let us apply the Erlang $\mathscr{B}$ formula to find m such that:

$$\mathscr{B}(m, A_0) \leq 0.05$$

In case of "medium" schools $\mathscr{B}(3, 1.2) = 0.09$ and $\mathscr{B}(4, 1.2) = 0.026$, so we obtain:

$$m_m = 4$$

Similarly, for "large" schools $\mathscr{B}(4, 2) = 0.095$ and $\mathscr{B}(5, 2) = 0.037$, then we obtain:

$$m_l = 5$$

Overall the call center service would require the deployment of $6m_m + 6m_l = 54$ operators.

If we now assume that the call centers are equipped with an automatic answering system that puts the calls on hold when all the operators are busy, then we must focus on the waiting time experienced by those callers that have to wait for an answer. The goal is to limit the waiting time below $\epsilon_0 \leq 3$ min for at least 95% of the callers. Therefore, we can use the formula of the probability distribution of ϵ and get:

$$F_\varepsilon(\varepsilon_0) = 1 - e^{-(m\mu-\lambda)\varepsilon_0} \geq \pi_0$$

that gives:

$$m\mu - \lambda \geq -\frac{\ln(1 - \pi_0)}{\varepsilon_0}$$

and then:

$$m \geq \frac{1}{\mu}\left(\lambda - \frac{\ln(1 - \pi_0)}{\varepsilon_0}\right) \qquad (4.88)$$

which is a different way to write Eq. (4.87). For "medium" and "large" schools we get, respectively:

$$m_m \geq 4.196 \rightarrow m_m = 5$$

$$m_l \geq 4.996 \rightarrow m_l = 5$$

In this case the same number of operators is sufficient for each call center of both kinds of schools. It is also worth noting that the probabilities of being queued are:

$$\pi_{rm} = \mathscr{C}(5, 1.2) = 8.21 \ 10^{-3}$$

$$\pi_{rl} = \mathscr{C}(5, 2) = 5.97 \ 10^{-2}$$

Of such small percentages of customers that have to wait, only 5% or less experience a waiting time in the queue longer than the target value. Therefore, this system is dimensioned in a very conservative way. In this case the overall service implementation would require 60 operators, a number that is a few units larger than in the previous case, but with the added value that almost no customers are not served as expected.

Finally, let us imagine a centralized call center. This would require some specific planning to guarantee operators that can answer questions regarding all schools, but concentrating all the effort should provide similar performance with less resources. The overall call arrival rate in this case is $\lambda = 6\lambda_m + 6\lambda_l = 6.4$ calls/min, corresponding to an overall offered traffic $A_0 = 6A_{0m} + 6A_{0l} = 19.2$ E. By applying formula (4.88) to the single centralized call center we obtain:

$$m \geq 22.196 \ \rightarrow \ m = 23$$

which is less than a half of the operators in the previous cases with dedicated call centers. The probability of being queued is:

$$\pi_r = \mathscr{C}(23, 19.2) = 0.31$$

In this case, many more customers end up being queued. This was expected and can be explained if we look at the servers utilization. For dedicated call centers the utilization is:

$$\rho_m = \frac{1.2}{5} = 0.24$$

$$\rho_l = \frac{2}{5} = 0.4$$

whereas for the centralized service the utilization is:

$$\rho = \frac{19.2}{23} = 0.835$$

A larger system makes a better use of the resources with a larger utilization, at the expenses of a larger probability of being queued, but with the same probability to wait more than a given target time.

4.3.4.2   Adopting a Unique Emergency Telephone Number

All countries have several emergency services devoted to different kinds of emergency: medical, public disorders, fires, environmental threats, etc. These emergency services are usually contacted by using some special telephone numbers, and different kinds of emergency are answered at different numbers. For instance, in Italy 118 is devoted to medical issues, 113 is used to call the police, 115 is used to call the firefighters, etc.

According to the European Universal Service Directive of 2002, article 26, a unique emergency number should be adopted. The "single European emergency call number" is 112 and is intended to connect the caller to a call center, where operators with a generic background evaluate the nature of the emergency and decide to which specific emergency service the caller must be connected. Since many calls to emergency services are inappropriate (e.g., requests for information, replicated calls for same event, etc.), the 112 operator may decide to simply dismiss a call, thus reducing the workload on the specific emergency operators and, hopefully, improving the overall quality of service.

The goal is to evaluate whether the 112 service leads to an actual improvement of the quality, or to a smaller number of operators needed to reach a given quality level. Therefore, we must to compare the quality achieved by the following two alternatives:

- A flat architecture with a set of call centers specialized per type of emergency, without the adoption of a unique number, as illustrated in Fig. 4.14. The traffic is routed to the specific call centers on the basis of the number dialed by the caller, and the specialized operators must deal with all calls, including the inappropriate ones.

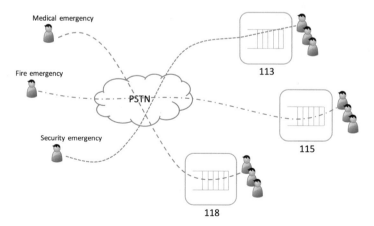

Fig. 4.14 Emergency services implemented with one dedicated call center for each specific emergency number

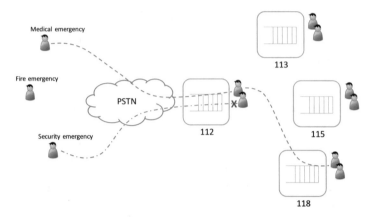

Fig. 4.15 Emergency services implemented with a hierarchy of call centers, adopting a unified emergency number. The operators of the first-level call center may decide to drop a call if deemed inappropriate and not forward it to the specialized call centers

- A hierarchical architecture with a centralized, first-level call center, where generic operators answer to all calls to the 112 service, and a set of specific, second-level call centers, with operators specialized per type of emergency, as shown in Fig. 4.15. Each call received by the 112 operators is either routed to the specialized operators or dismissed if found inappropriate.

The traffic assumptions during peak hours are that the average emergency call arrival rate is $\lambda = 900$ calls/h, that the time a specialized operator needs to deal with a call is $\bar{\vartheta} = 3$ min, and that $\pi_i = 0.4$ is the probability that a call is inappropriate and can be dismissed. The quality goal to achieve is that a caller that finds all the operators busy waits on average for less than $\epsilon_0 = 2$ min.

Let us consider the first alternative without single emergency number, assuming to have 3 specialized call centers and that the traffic is evenly spread among them, i.e., the call arrival rate to each call center is $\lambda/3 = 300$ calls/h. The offered traffic per call center is:

$$A_0 = \frac{\lambda}{3}\bar{\vartheta} = \frac{300}{60} \cdot 3 = 15\,\text{E}$$

Knowing the expression of $\bar{\epsilon}$ we can impose:

$$\bar{\epsilon} = \frac{\bar{\vartheta}}{m - A_0} \leq \epsilon_0$$

and then obtain:

$$m \geq A_0 + \frac{\bar{\vartheta}}{\epsilon_0} = 16.5$$

Overall, the flat architecture requires at least $m = 17$ specialized operators for each of the three call centers, for a total of 51 operators.

Let us now consider the second alternative, with a single emergency number and the hierarchical architecture. We need to make some additional assumptions to fully solve the case. Let us assume that the 112 operators need far less time to process the call and decide whether to route it further or dismiss it, say $\bar{\vartheta}' = 0.3\,\text{min}$, and that customers that have to wait must not wait in the 112 queue for more than $\epsilon_0' = 0.5\,\text{min}$. Under these assumptions, the traffic offered to the first-level call center is:

$$A_0' = \lambda\bar{\vartheta}' = \frac{900}{60}\cdot 0.3 = 4.5\,\text{E}$$

It follows that:

$$m' \geq A_0' + \frac{\bar{\vartheta}'}{\epsilon_0'} = 5.1$$

Therefore, at least $m' = 6$ generic operators are needed in the first-level call center.

As for the second-level call centers, the incoming traffic is reduced to 60% of the original value, thanks to the filtering made by the 112 operators. Moreover, let us impose $\epsilon_0 = 1.5\,\text{min}$ to compensate for the waiting time at the first-level call center. We have:

$$A_0 = \frac{\lambda}{3}(1 - \pi_i)\bar{\vartheta} = \frac{300\cdot 0.6}{60}\cdot 3 = 9\,\text{E}$$

and therefore:

$$m \geq A_0 + \frac{\bar{\vartheta}}{\epsilon_0} = 11$$

In summary, we need at least $3 \times m = 33$ specialized operators, to whom we have to add the $m' = 6$ generic operators of the centralized service, for a total of 39 operators. With the adoption of a single emergency number and a centralized management of the emergency calls, we achieve interesting resource savings with respect to the 51 operators in the case of completely separated services.

4.3.4.3 Planning Lines and Operators in a Call Center

A company is deploying a call center to offer customer support. The goal is to plan the resources to be allocated to the call center to guarantee operations over the next 3 years. At first, the company made some market analysis to forecast the traffic demand to the call center. The conclusion is that the peak traffic at startup is estimated to be $A_0(0) = 12\,\text{E}$, with an annual increase of 20%. The calls arriving

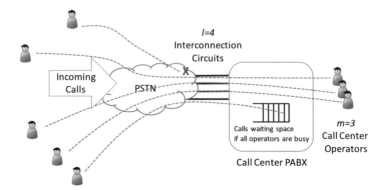

Fig. 4.16 Scheme of the call center system. In this example, $l = 4$ circuits connect the call center with the PSTN and $m = 3$ operators are present in the call center. In this simple example it may happen that 3 users are connected to the call center and a fourth one calls in, which is queued waiting for one of the operators to get free

at the call center are assumed to be randomly distributed according to a Poisson process and with an exponential random call holding time with average $\bar{\vartheta} = 3$ min.

For budget planning reasons, the company plans to implement upgrades to the call center in terms of resources at the beginning of each year. The company management needs to know the planning of the total number of resources required in the call center, including:

- the number l of circuits interconnecting the call center with the PSTN to collect customers calls; l must be large enough to guarantee a call blocking probability $\pi_p \leq 0.01$;
- the number m of call center operators to guarantee a waiting time for the customers that have to wait that is less than $\epsilon_0 = 1$ min with probability $\pi_0 \geq 0.9$.

With respect to the previous case studies, where the focus was only on the number of operators in the call center, here we have to consider also the number of circuits interconnecting the call center with the PSTN, as illustrated in Fig. 4.16. This is an important issue and when dealing with it we face three possibilities:

- If $l < m$, exactly $m - l$ operators are completely useless. They sit at the call center but will never receive a call, since only up to l users can reach the call center from the PSTN.
- If $l = m$, the call center operates as a pure loss system. Calls will never wait in the queue. If they find a free operator they are immediately answered, otherwise they are blocked due to lack of circuits. We already know that the blocking probability is tuned by choosing the right number of servers, but we also know that the sever utilization of this sort of systems is rather low in general.
- If $l > m$, calls can enter the call center even when all operators are busy. In this case the call center PABX may be equipped with waiting space (typically letting

the caller listening to some music or prerecorded messages), and this is the case we have to consider here. We know that the utilization can be increased by adding a waiting space, and this is the scope of this solution.

Now let us see how the dimensioning of this call center can be achieved, given the performance goals mentioned above. Let us start by finding the incoming traffic. Knowing that the traffic increases by 20% each year, we can write that at the end of the first year:

$$A_0(1) = A_0(0)e^{\frac{1}{\tau}} = 12e^{\frac{1}{\tau}} = 14.4 \text{ E}$$

and then:

$$\tau = \frac{1}{\ln(1.2)} = 5.485 \text{ years}$$

Therefore, the expression of the incoming traffic as a function of time is:

$$A_0(t) = A_0(0)e^{\frac{t}{5.49}}$$

and the values at the end of each year are:

$$A_0(1) = 14.4 \text{ E}$$

$$A_0(2) = 12e^{\frac{2}{5.49}} = 17.3 \text{ E}$$

$$A_0(3) = 12e^{\frac{3}{5.49}} = 20.8 \text{ E}$$

The average arrival rates corresponding to these values, considered as offered traffic, are:

$$\lambda(0) = \mu A_0(0) = \frac{12}{3} = 4 \text{ calls/min}$$

$$\lambda(1) = \mu A_0(1) = \frac{14.4}{3} = 4.8 \text{ calls/min}$$

$$\lambda(2) = \mu A_0(2) = \frac{17.3}{3} = 5.8 \text{ calls/min}$$

$$\lambda(3) = \mu A_0(3) = \frac{20.8}{3} = 6.9 \text{ calls/min}$$

Now we have two problems to solve that are correlated.

- To determine l, we need to apply the Erlang $\mathscr{B}$ formula. But we do not know the actual value of the offered traffic, because the call holding time from the point of view of the PSTN circuits is given by $\bar{\vartheta}_{PSTN} = \bar{\eta} + \bar{\vartheta}$. In words, the call holding time, which keeps the circuits busy, is the sum of the time spent by the caller waiting for an operator and the time spent talking to the operator. If η and ϑ are exponentially distributed, their sum is not exponential anymore, but we already know that this is not an issue thanks to the insensitivity of the Erlang $\mathscr{B}$ formula to the service time distribution.
- To determine m we must apply the formula of the waiting time in the $\mathscr{M}/\mathscr{M}/m$ system, but again we do not know the exact value of the offered traffic, since a percentage of the calls are blocked.

In principle, finding an exact solution of these two problems at once is rather complex. However, we can use a simple approximation with good effectiveness when $\pi_p \ll 1$, which should be the case when targeting a reasonable quality of service. In this case the traffic lost because of lack of circuits with the PSTN is small enough to be considered negligible, i.e., $A_p \simeq 0$. Then we can calculate the number of operators considering the distribution of the waiting time for customers that wait:

$$F_\varepsilon(t) = 1 - e^{-(m\mu - \lambda)t}$$

and imposing that $F_\varepsilon(\varepsilon_0) \geq \pi_0$, we obtain:

$$m \geq \bar{\vartheta}\left(\lambda - \frac{1}{\epsilon_0}\ln(1 - \pi_0)\right)$$

Using this formula we can take into account the time variation of the traffic, leading to:

$$m(t) \geq \bar{\vartheta}(\lambda(t+1) - \ln 0.1) = 3(\lambda(t+1) + 2.3)$$

because the upgrade made at the beginning of each year must guarantee the required quality level until the end of the year. Then we can plan the number of operators at the beginning of each year of operations:

$$m(0) \geq 3(\lambda(1) + 2.3) = 3(4.8 + 2.3) = 21.3 \quad \rightarrow \quad m(0) = 22$$

$$m(1) \geq 3(\lambda(2) + 2.3) = 3(5.8 + 2.3) = 24.3 \quad \rightarrow \quad m(1) = 25$$

$$m(2) \geq 3(\lambda(3) + 2.3) = 3(6.9 + 2.3) = 27.6 \quad \rightarrow \quad m(2) = 28$$

Once m is known we can calculate $\bar{\eta}$:

$$\bar{\eta}(t) = \mathscr{C}\,(m(t), A_0(t))\,\frac{\bar{\vartheta}}{m(t) - A_0(t)}$$

and, with reference to the end of each year of operations when the traffic is maximum with respect to the number of operators available, we get:

$$\bar{\eta}(1) = 0.0435\frac{3}{7.6} = 0.017 \text{ min}$$

$$\bar{\eta}(2) = 0.057\frac{3}{7.7} = 0.022 \text{ min}$$

$$\bar{\eta}(3) = 0.093\frac{3}{7.2} = 0.039 \text{ min}$$

and:

$$\bar{\vartheta}_{PSTN}(1) = \bar{\vartheta} + \bar{\eta} = 3.017 \text{ min}$$

$$\bar{\vartheta}_{PSTN}(2) = 3.022 \text{ min}$$

$$\bar{\vartheta}_{PSTN}(3) = 3.039 \text{ min}$$

It is easy now to get the approximate values of the actual offered traffic:

$$A_0'(1) = \lambda(1)\bar{\vartheta}_{PSTN}(1) = 4.8 \cdot 3.017 = 14.48$$

$$A_0'(2) = \lambda(2)\bar{\vartheta}_{PSTN}(2) = 5.8 \cdot 3.022 = 17.52$$

$$A_0'(3) = \lambda(3)\bar{\vartheta}_{PSTN}(3) = 6.9 \cdot 3.039 = 20.97$$

and the number of circuits l needed at the beginning of each year of operations:

$$l(0) = 24$$

$$l(1) = 27$$

$$l(2) = 31$$

It is important to outline that this approximate solution method is conservative. We slightly over-estimate the traffic at the operators and, consequently, the waiting time and the number of circuits to connect to the PSTN. Indeed, if π_p is higher the approximation may be quite rough. In that case, it is possible to refine the solution by iterating the approximated computation. With the values of l obtained at a previous iteration, we can calculate the lost traffic and solve for m with lesser traffic to get a better estimate. Then we find new values of l and repeat the computation. This procedure usually converges to rather stable values in a few iterations.

4.3.4.4 A Call Center with Impatient Customers

A call center must provide services to a pool of customers generating random calls according to a Poisson process with $\lambda = 30$ calls/h with exponential call holding time with average $\bar{\vartheta} = 2$ min. When all operators are busy, the customers are put on hold in a FIFO queue. However, customers are impatient and decide to hang up the call when their waiting time becomes longer than a certain value ζ. Let us assume that such "patience" time is also distributed according to an exponential distribution, with average $\bar{\zeta} = \frac{1}{\mu_a}$, and that all customers behave independently one another.

The system can be described as a BD process with:

$$\lambda_k = \lambda \quad \forall k \geq 0$$

$$\mu_k = \begin{cases} k\mu & 0 < k \leq m \\ m\mu + (k-m)\mu_a & \forall k > m \end{cases} \tag{4.89}$$

This is also called the Palm/Erlang-A model, or $\mathcal{M}/\mathcal{M}/m + I$ model. We can write the steady state probabilities as:

$$P_k = \begin{cases} P_0 \frac{A_0^k}{k!} & 0 \leq k \leq m \\ P_0 \frac{A_0^m}{m!} \prod_{i=m+1}^{k} \frac{\lambda}{m\mu+(i-m)\mu_a} = P_0 \frac{A_0^k}{m!} \prod_{i=m+1}^{k} \frac{1}{m+(i-m)\frac{\mu_a}{\mu}} & \forall k > m \end{cases} \tag{4.90}$$

with:

$$P_0 = \left[\sum_{k=0}^{m} \frac{A_0^k}{k!} + \sum_{k=m+1}^{\infty} \frac{A_0^k}{m!} \prod_{i=m+1}^{k} \frac{1}{m + (i-m)\frac{\mu_a}{\mu}} \right]^{-1} \tag{4.91}$$

As in other cases when the queuing space is unlimited, we should understand the conditions that make the system stable. In other words, we have to verify whether the series in (4.91) converges. For any state with at least one customer in the queue, recalling (4.89) we can write the following inequality:

$$k \min(\mu, \mu_a) \leq \mu_k \leq k \max(\mu, \mu_a) \quad \forall k > m \tag{4.92}$$

which can be rewritten as:

$$\frac{k \min(\mu, \mu_a)}{\mu} \le m + (k - m)\frac{\mu_a}{\mu} \le \frac{k \max(\mu, \mu_a)}{\mu} \quad \forall k > m \tag{4.93}$$

where the equal signs apply when $\mu_a = \mu$. Considering that, obviously, $\mu \ge \mu_{\min} = \min(\mu, \mu_a)$, we can find an upper bound to the reciprocal of P_0 as follows:

$$
\begin{aligned}
P_0^{-1} &= \sum_{k=0}^{m} \frac{A_0^k}{k!} + \sum_{k=m+1}^{\infty} \frac{A_0^k}{m!} \prod_{i=m+1}^{k} \frac{1}{m + (i - m)\frac{\mu_a}{\mu}} \\
&\le \sum_{k=0}^{m} \frac{A_0^k}{k!} + \sum_{k=m+1}^{\infty} \frac{A_0^k}{m!} \prod_{i=m+1}^{k} \frac{\mu}{i \, \mu_{\min}} = \sum_{k=0}^{m} \frac{A_0^k}{k!} + \sum_{k=m+1}^{\infty} \frac{A_0^k}{m!} \frac{\mu^{k-m}}{\frac{k!}{m!} \mu_{\min}^{k-m}} \\
&= \sum_{k=0}^{m} \frac{(\lambda/\mu)^k}{k!} + \sum_{k=m+1}^{\infty} \frac{(\lambda/\mu)^m}{k!} \frac{\lambda^{k-m}}{\mu_{\min}^{k-m}} \le \sum_{k=0}^{m} \frac{(\lambda/\mu_{\min})^k}{k!} + \sum_{k=m+1}^{\infty} \frac{(\lambda/\mu_{\min})^k}{k!} \\
&= \sum_{k=0}^{\infty} \frac{(\lambda/\mu_{\min})^k}{k!} = e^{\lambda/\mu_{\min}}
\end{aligned}
\tag{4.94}
$$

The above inequality is valid for any positive value of λ, μ, and μ_a, which means that $P_0 > 0$ for any value of A_0. Therefore, the BD process is always stable, independently of the offered load value. This can be easily justified by the fact that the higher the number of customers in the queue, the higher the frequency with which a customer abandons the queue, as confirmed by the $(k - m)\mu_a$ term in (4.89). This implies that, when $\rho > 1$, the growing number of abandoning events compensate the growing number of arrivals, and the number of queued customers never grows to infinity.

Similarly to the $\mathcal{M}/\mathcal{M}/m$ system, we can define the congestion probability as the probability that a customer has to wait. Thanks to the PASTA property, this probability is given by the sum of the state probabilities for all the states where all servers are busy:

$$\pi_r = \Pr\{k \ge m\} = \sum_{k=m}^{\infty} P_k = \sum_{k=m}^{\infty} P_0 \frac{A_0^k}{m!} \prod_{i=m+1}^{k} \frac{1}{m + (i - m)\frac{\mu_a}{\mu}} = \mathcal{A}(m, A_0, \mu_a) \tag{4.95}$$

which is often called the Erlang $\mathcal{A}$ formula.

The steady state probabilities in (4.90) and (4.91) are not as easy to calculate as the ones of the system without impatient customers, but they can be obtained numerically. However, if we assume that $\mu_a = \mu$, the steady state probabilities become the same as in a $\mathcal{M}/\mathcal{M}$ system with infinite servers, as obtained in Sect. 4.2.2.

Now let us assume that the system is equipped with a single operator, i.e., $m = 1$. We are interested in:

1. the probability that a customer has to wait when calling the call center;
2. the average number A_c of customers in the queue;
3. the throughput A_s and the utilization of the servers (operators);
4. the percentage of lost customers, i.e., the percentage of customers that leave the system without being served because of their impatience.

First, let us find the offered traffic:

$$A_0 = \lambda \bar{\vartheta} = 1 \, \text{E}$$

This means that $A_0 = m$, but it does not pose any problem because the impatience of the customers guarantees that the system is stable. The probability that a customer has to wait is:

$$\pi_r = \sum_{k=m}^{\infty} P_k = 1 - \sum_{k=0}^{m-1} P_k = 1 - \sum_{k=0}^{m-1} \frac{A_0^k}{k!} e^{-A_0} = \mathscr{A}(m, A_0)$$

and for $m = 1$:

$$\pi_r = 1 - P_0 = 1 - e^{-A_0} = 0.632$$

The average number of customers in the queue can be calculated by definition from the steady state probabilities given that, when k customers are in the system with $k \geq m$, then $h = k - m$ customers are in the queue:

$$A_c = \sum_{k=m}^{\infty} (k - m) P_k$$

and for $m = 1$:

$$A_c = \sum_{k=1}^{\infty} (k-1) P_k = \sum_{k=1}^{\infty} k P_k - \sum_{k=1}^{\infty} P_k = \sum_{k=1}^{\infty} k \frac{A_0^k}{k!} P_0 - (1 - P_0)$$

$$= A_0 \sum_{k=1}^{\infty} \frac{A_0^{k-1}}{(k-1)!} e^{-A_0} - (1 - e^{-A_0}) = A_0 - 1 + e^{-A_0} = 0.368$$

The throughput of the system is given by the average number of customer served and can be calculated noting that the number of served customers is equal to the number of customers in the system when $k \leq m$, and is m for all other values of k:

$$A_s = \sum_{k=0}^{m-1} k P_k + \sum_{k=m}^{\infty} m P_k$$

With $m = 1$:

$$A_s = \sum_{k=1}^{\infty} P_k = 1 - P_0 = 1 - e^{-A_0} = 0.632$$

The server utilization is given by the percentage of time the server is busy. When $m = 1$:

$$\rho = \sum_{k=1}^{\infty} P_k = 1 - P_0 = 1 - e^{-A_0} = 0.632$$

Finally, the percentage of lost customers is the ratio between the lost traffic (difference between the offered traffic and the throughput) and the offered traffic:

$$\pi_p = \frac{A_0 - A_s}{A_0} = 0.368$$

4.4   Multi-Dimensional BD Processes

In the previous sections we assumed that all customers sending call requests to a circuit switching system were identical. In a real environment, however, there may be cases in which different customer streams have different characteristics. So they must be accounted for separately when considering the overall behavior of the system. In the examples that follow we provide some reasons for this, but for the time being let us simply assume that we have to keep track of the number of each type of customers in the system.

Let us then index the n customer groups from 1 to n, and let us consider a generic circuit switching system without waiting space. We define:

- k_i as the number of customers of type i in the system, with $i = 1, 2, \ldots, n$;
- $\mathbf{k} = (k_1, k_2, \ldots, k_n)$ as the state of the system, expressed as a vector of size n with the number of customers of each type currently in the system.

A multi-dimensional birth–death (mBD) process is defined as follows:

- arrival and service processes are memoryless, therefore the system can be modeled as a continuous time Markov Chain;
- the state of the system changes by one customer at a time, i.e., if $\mathbf{e}_i = (0, 0, \ldots, 0, 1, 0, \ldots, 0)$ is the vector with all elements equal to 0 except the i-th which is equal to 1, then the only state transitions allowed in a mBD process are:

  – from $\mathbf{k}$ to $\mathbf{k} + \mathbf{e}_i$ when a customer of type i arrives
  – from $\mathbf{k}$ to $\mathbf{k} - \mathbf{e}_i$ when a customer of type i leaves

  for any $i = 1, 2, \ldots, n$.

If all the arrival processes of the n customer types are Poisson processes, then λ_i is the arrival rate of type i customers and the overall arrival process is also a Poisson process resulting from the composition of the n processes. If the call holding time (or service time) of any type i customers is exponentially distributed with average $\bar{\vartheta}_i$ and does not depend on other customers in service, then $\mu_i(k_i)$ is the service rate of type i customers when k_i of them are present in the system.

In general, $\mathbf{k}$ takes values in a so-called *state space* $\mathscr{S}_0$. Let us say that $k_i \in \mathscr{S}_i \ \forall i$, where $\mathscr{S}_i$ is the set of possible values for k_i. If the k_i-s are independent and do not influence one another, then the state space of the system is given by the cartesian product of each $\mathscr{S}_i$, i.e., $\mathbf{k} \in \mathscr{S}_0 = \mathscr{S}_1 \times \mathscr{S}_2 \times \ldots \times \mathscr{S}_n$. In practice, this means that any type of customer behaves independently of the others and the mBD process is simply the combination of n independent BD processes, each modeled as an $\mathcal{M}/\mathcal{M}/m/0$ system as discussed in Sect. 4.2. We can then calculate the steady state probabilities for the BD process of type i customers as P_{ki}. Since the processes are independent, the overall system steady state probability $P_{\mathbf{k}}$ is given by the joint probability of the single BD process states:

$$P_{\mathbf{k}} = P_{k1} \cdot P_{k2} \cdot \ldots \cdot P_{kn} = \prod_{i=1}^{n} P_{0i} \left(\frac{\lambda_i}{\mu_i} \right)^{k_i} \frac{1}{k_i!} = \prod_{i=1}^{n} P_{0i} \frac{A_{0i}^{k_i}}{k_i!} \tag{4.96}$$

This is the most straightforward case but, in general, it may happen that k_i depends in some way on k_j, with $j \neq i$. It follows that some states in $\mathscr{S}_0$ may not be feasible and the state space of the system is restricted to some $\mathscr{S} \subset \mathscr{S}_0$. This is usually called a *truncated* state space, meaning that some of the elements in $\mathscr{S}_0$ are cut off because the related combination of customers cannot happen.

As a very easy and intuitive example to understand this concept, let us consider a system with $n = 2$ customer types and $m = 2$ servers. Obviously, if we consider any type of customers independently, we have $0 \leq k_1 \leq 2$ and $0 \leq k_2 \leq 2$. Therefore, $\mathscr{S}_1 = \{0, 1, 2\}$ and $\mathscr{S}_2 = \{0, 1, 2\}$. As a consequence, the state space is:

$$\mathscr{S}_0 = \mathscr{S}_1 \times \mathscr{S}_2 = \{(0,0), (0,1), (0,2), (1,0), (1,1), (1,2), (2,0), (2,1), (2,2)\}$$

However, $m = 2$ limits the overall number of customers that can be in the system. It follows that k_1 and k_2 are subject to the joint constraint $k_1 + k_2 \leq 2$. Therefore, the state of the system can take values only in the set:

$$\mathscr{S} = \{(0,0), (0,1), (0,2), (1,0), (1,1), (2,0)\}$$

and clearly $\mathscr{S} \subset \mathscr{S}_0$.

The constraint we have used in this example is given by a linear expression and can be generalized as follows. Given:

- a matrix $\mathbf{B}$ of size $n \times h$;
- a vector $\mathbf{C}$ of size h;

then the formula

$$\mathbf{k} \, \mathbf{B} \leq \mathbf{C} \tag{4.97}$$

defines a new state space $\mathscr{S}$ which is called *a linear restriction* of state space $\mathscr{S}_0$. In the examples with $n = 2$ and $m = 2$, the state space restriction is $k_1 + k_2 \leq 2$, therefore $h = 1$, $\mathbf{B} = (1, 1)^T$, $\mathbf{C} = 2$.

Now the question is: what happens to the steady state probabilities when the state space is truncated? The general answer is that we must start from scratch and analyze the truncated system as a new system. Fortunately, for a mBD process this is not the case. It is possible to prove that such process is *reversible*. It happens that, if we take a reversible Markov process with state space $\mathscr{S}_0$ and steady state probabilities $P_\mathbf{k}$ and imagine to truncate the state space to $\mathscr{S} \subset \mathscr{S}_0$, the new steady state probabilities $P'_\mathbf{k}$ can be obtained from $P_\mathbf{k}$ by simply recalculating the normalizing constant P_0.

Therefore, for the queuing system without waiting space we are considering here:

$$P'_\mathbf{k} = \begin{cases} G(\mathscr{S})^{-1} \prod_{i=1}^{n} \frac{A_{0i}^{k_i}}{k_i!} & \mathbf{k} \in \mathscr{S} \\ \\ 0 & \mathbf{k} \notin \mathscr{S} \end{cases} \tag{4.98}$$

where $G(\mathscr{S})$ has the following expression:

$$G(\mathscr{S}) = \sum_{\mathbf{k} \in \mathscr{S}} \left(\prod_{i=1}^{n} \frac{A_{0i}^{k_i}}{k_i!} \right) \tag{4.99}$$

In conclusion, we can analyze the mBD process very easily. For a circuit switching system without queuing space we know the expression of the steady state probabilities from Eq. (4.96) and we have to calculate $G(\mathscr{S})$. This is rather simple in principle with reference to Eq. (4.99), but can be very demanding in terms of computation time, if the state space is large.

We still have to find how to obtain the performance of the system in terms of call blocking probability for the different customer types. To this end, let us give with some definitions.

Let us define $\mathscr{S}_{bi} \subset \mathscr{S}$ as the set of states such that, if a new call request of type i arrives, it cannot be accepted and is blocked, i.e.:

$$\mathscr{S}_{bi} = \{\mathbf{k} \in \mathscr{S} : \mathbf{k} + \mathbf{e_i} \notin \mathscr{S}\} \tag{4.100}$$

Let us call $\overline{\mathscr{S}}_{bi}$ the complement of $\mathscr{S}_{bi}$ with respect to $\mathscr{S}$, i.e.:

$$\mathscr{S}_{bi} \cup \overline{\mathscr{S}}_{bi} = \mathscr{S} \quad \text{and} \quad \mathscr{S}_{bi} \cap \overline{\mathscr{S}}_{bi} = \varnothing \tag{4.101}$$

Owing to the PASTA property, the blocking probability for customers of type i is the sum of the steady state probabilities of states where the arrival of a customer of type i cannot be accommodated in the system. It follows that:

$$\pi_{pi} = \sum_{\mathbf{k} \in \mathscr{S}_{bi}} P_{\mathbf{k}} = G(\mathscr{S})^{-1} \sum_{\mathbf{k} \in \mathscr{S}_{bi}} \left(\prod_{i=1}^{n} \frac{A_{0i}^{k_i}}{k_i!} \right) = \frac{G(\mathscr{S}_{bi})}{G(\mathscr{S})} \tag{4.102}$$

where $G(\mathscr{S}_{bi})$ is the normalizing constant as defined in (4.99) but restricted to state subspace $\mathscr{S}_{bi}$. From Eq. (4.99) and the definition of $\overline{\mathscr{S}}_{bi}$ it follows that:

$$G(\mathscr{S}) = G(\mathscr{S}_{bi}) + G(\overline{\mathscr{S}}_{bi}) \tag{4.103}$$

and then:

$$\pi_{pi} = 1 - \frac{G(\overline{\mathscr{S}}_{bi})}{G(\mathscr{S})} \tag{4.104}$$

Indeed $\frac{G(\overline{\mathscr{S}}_{bi})}{G(\mathscr{S})} \leq 1$ and $0 \leq \pi_{pi} \leq 1$.

In summary, we have shown that the analysis of a mBD system is related to our capability to calculate two normalizing constants over two different state spaces. The formulas to do so are known but the solution needs numerical analysis. As outlined above, the computation may be challenging for very large state spaces. Specific techniques are available to mitigate the computation complexity problem, but we do not discuss them here. The problems we address in the following sections are all easily treatable with state-of-the-art computers.

Last but not least, a useful parameter can be defined as follows:

The *normalized offered load* of traffic class i is given by the quantity:

$$a_i = A_{0i} \frac{b_i}{C} \tag{4.105}$$

where A_{0i} is the offered load of traffic class i, b_i is the amount of server capacity requested by a class i customer, and C is the total server capacity.

The normalized offered load is a very useful parameter. If $a_i = a_j$ with $i, j = 1, \ldots, n$ and $i \neq j$, then traffic classes i and j require the same average capacity per time unit, even if they have different absolute values of the offered load. For instance, if class i is less demanding in terms of capacity, i.e., $b_i < b_j$, its offered load must be higher, i.e., $A_{0i} > A_{0j}$, in order to balance the average required capacity. Therefore, by considering the normalized offered load we can compare in a sort of fair resource request conditions the behavior of the different traffic classes in terms of QoS.

4.4.1   The Multi-Service Link

Let us consider the system depicted in Fig. 4.17. A link with capacity (bandwidth or bit rate) C is shared among n classes of customers with different capacity requirements. Each customer of class i requires a bandwidth (or bit rate) equal to b_i. The system works according to a circuit switching principle without queuing. When a call request of class i arrives, if it can be accommodated because there is enough free capacity, then it is accepted, otherwise it is blocked. Let us assume that calls of class i arrive according to a Poisson process with average arrival rate λ_i and that their call holding time has exponential distribution with average $\bar{\vartheta}_i = 1/\mu_i$. Let us represent the capacity requirements of all traffic classes with the vector $\mathbf{b} = (b_1, \ldots, b_i, \ldots b_n)$.

If C is infinite, the system is able to accept any request by any customer of any class. The state space is $\mathscr{S}_0 = \mathscr{S}_1 \times \mathscr{S}_2 \times \ldots \times \mathscr{S}_n$ with $\mathscr{S}_i = \{0, 1, 2, \therefore, \infty\}$. There are no joint bounds between the values of the k_i-s and any type of customer can be studied independently with Eq. (4.96). On the other hand, when C is finite there is obviously a limit on the total number of calls that can be accepted, which can be expressed as:

$$\sum_{i=1}^{n} k_i b_i \leq C \qquad (4.106)$$

In other words, if $C_r = C - \sum_{i=1}^{n} k_i b_i$ is the residual link capacity, when a new call of class i arrives it can be accepted if and only if $b_i \leq C_r$. Otherwise it is blocked. In general, customers of class i experience a blocking probability given by π_{pi}.

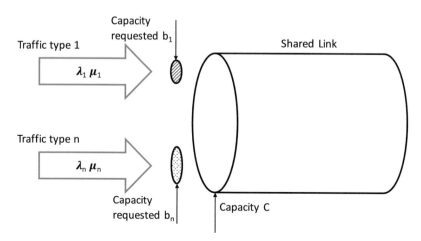

Fig. 4.17 A link with capacity C is shared by n types of customers with different capacity requirements

In this case the number of customers of the different traffic classes that can be accepted in the system are not independent anymore. The state space $\mathscr{S}_0$ is thus truncated, and the constraint in Eq. (4.106) defines $\mathscr{S}$ as a linear restriction of $\mathscr{S}_0$, according to the definition given in (4.97). In particular, $\mathbf{B}$ is a $n \times 1$ matrix, i.e.:

$$\mathbf{B} = \begin{pmatrix} b_1 \\ b_2 \\ \vdots \\ b_n \end{pmatrix} \tag{4.107}$$

and $\mathbf{C} = C$ is a scalar. Then the relevant state spaces are:

$$\mathscr{S} = \{\mathbf{k} \mid \sum_{i=1}^{n} k_i \, b_i \leq C\} \tag{4.108}$$

$$\overline{\mathscr{S}_{bi}} = \{\mathbf{k} \mid \sum_{i=1}^{n} k_i \, b_i \leq C - b_i\} \tag{4.109}$$

and the blocking probability for traffic class i is:

$$\pi_{pi} = 1 - \frac{G(\overline{\mathscr{S}_{bi}})}{G(\mathscr{S})} \tag{4.110}$$

where:

$$\begin{aligned} G(\overline{\mathscr{S}_{bi}}) &= \sum_{\mathbf{k} \mid \sum_{i=1}^{n} k_i \, b_i \leq C - b_i} \left(\prod_{i=1}^{n} \frac{A_{0i}^{k_i}}{k_i!} \right) \\ G(\mathscr{S}) &= \sum_{\mathbf{k} \mid \sum_{i=1}^{n} k_i \, b_i \leq C} \left(\prod_{i=1}^{n} \frac{A_{0i}^{k_i}}{k_i!} \right) \end{aligned} \tag{4.111}$$

It is easy to show that, if we take two traffic classes i and j such that $b_i \geq b_j$, then $\pi_{pi} \geq \pi_{pj}$. Indeed. since $C - b_i \leq C - b_j$, it follows that $\overline{\mathscr{S}_{bi}}$ includes the same or less states than $\overline{\mathscr{S}_{bj}}$, therefore $G(\overline{\mathscr{S}_{bi}}) \leq G(\overline{\mathscr{S}_{bj}})$ and $\pi_{pi} \geq \pi_{pj}$.

4.4.2   Circuit-Switched Networks with Fixed Routing

Let us consider a circuit-switched network, i.e., a whole infrastructure with several central offices and trunk groups interconnecting them. The central offices in the network can be considered, without loss of generality, access nodes to which the end users are connected by local loops. Typically, customers do not share the local loop and therefore can reach the access node without congestion issues. However, they share the trunk groups between the nodes when they call customers that access the network from other nodes.

The incoming calls require a circuit on each trunk group needed to connect the source to the destination, according to a given network path or *route*.

A *route* is the set of circuits from the trunk groups that a call has to traverse from the source node i to the destination node j.

We assume that routes are pre-defined and cannot be changed, as explained below. Obviously, a new call can be accepted and set up correctly if and only if the required circuits along the route are available. When the calls arrive according to a Poisson process and the service time is exponentially distributed, this problem can be described as a mBD.

Now let us assume that the network operates according to the following principles:

1. the network has N nodes that may or may not be directly connected;
2. the network is equipped with J trunk groups that determine the network topology, and trunk group j has capacity C_j (i.e., it can accommodate C_j circuits), with $j = 1, 2, \ldots, J$;
3. R routes are possible, with $R \leq \frac{N(N+1)}{2}$, and calls requesting route r arrive as a Poisson process with rate λ_r, with $r = 1, 2, \ldots, R$;
4. a call on route r requires b_{rj} circuits on trunk group j, assuming that $b_{rj} = 1$ if j is part of the route r, while $b_{rj} = 0$ if j is not part of the route;
5. no alternative or dynamic routing is possible: the path between node X and node Y is predetermined by the routing algorithm and does not change in time;
6. the average call holding time is the same for all calls (regardless of the route they belong to) and is equal to $\bar{\vartheta} = \frac{1}{\mu}$.

As usual, if we let k_r be the number of calls using route r, the state of the system is given by $\mathbf{k} = (k_1, k_2, \ldots, k_R)$. The system can be described as a mBD process. Indeed the state space is truncated, since the number of calls on every route is limited by the fact that the sum of the capacity requests on a trunk group i cannot exceed the capacity C_i of the group, as shown in the example of Fig. 4.18. The matrix $\mathbf{B}$ is a $R \times J$ matrix:

$$\mathbf{B} = \begin{pmatrix} b_{11} & b_{12} & \cdots & b_{1J} \\ b_{21} & b_{22} & \cdots & b_{2J} \\ & & \vdots & \\ b_{R1} & b_{R2} & \cdots & b_{RJ} \end{pmatrix} \tag{4.112}$$

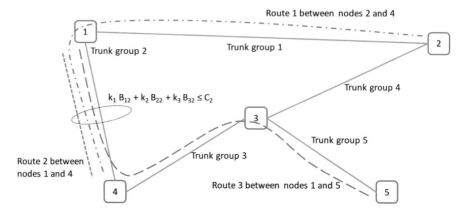

Fig. 4.18 An example of a circuit-switched network with 3 active routes: route 1 between nodes 2 and 4, route 2 between nodes 1 and 4, and route 3 between nodes 1 and 5. The three routes traverse the trunk group between nodes 1 and 4 and thus they share its capacity. This results in a constraint on the state space

and the vector **C** is:

$$\mathbf{C} = \begin{pmatrix} C_1 \\ C_2 \\ \vdots \\ C_J \end{pmatrix} \tag{4.113}$$

Finally the state space $\mathscr{S}$ is defined as usual as a linear restriction of $\mathscr{S}_0$ according to formula (4.97). In this case the constraint means that on each trunk group j the overall capacity usage must be less than the capacity available. The total capacity used on trunk group j is the sum of the capacity used by all the calls of every route traversing that trunk group:

$$\sum_{r=1}^{R} k_r \, b_{rj} \leq C_j \quad j = 1, 2, \ldots, J \tag{4.114}$$

Similarly, we can calculate the state space:

$$\overline{\mathscr{S}_{bj}} = \{\mathbf{k} \mid \mathbf{k}\mathbf{B} \leq \mathbf{C} - \mathbf{e}_j\mathbf{C}\} \tag{4.115}$$

and the related normalizing constants, as well as the blocking probability per route, using Eq. (4.104).

4.4.3 Examples and Case Studies

4.4.3.1 QoS Comparison for Two Traffic Classes with Different Bandwidth Requirements

A digital link of total capacity C carries calls belonging to two traffic classes with different bandwidth requirements b_1 and b_2. Let k_1 and k_2 be the number of active calls of type 1 and 2, respectively. $\mathbf{k} = \{k_1, k_2\}$ is the state of the system. We want to analyze the blocking probability π_1 and π_2 of traffic classes 1 and 2, respectively, for a total offered load ranging from 0 to 1 and evenly split between the two traffic classes, i.e., $a_1 = a_2$. As already outlined above, an equal normalized offered load assures that the average bandwidth demand is the same for the two traffic classes, meaning that calls requiring larger bandwidth are less frequent than calls requiring smaller bandwidth. The goal is to analyze the blocking probability in two different cases:

- case 1: two traffic classes with different bandwidth requests, namely $b_1 = 0.01\,C$ and $b_2 = 0.1\,C$;
- case 2: two traffic classes with very different bandwidth requests, namely $b_1 = 0.01\,C$ and $b_2 = 0.25\,C$.

The results of the computation of formula (4.110) are plotted in Figs. 4.19 and 4.20 for cases 1 and 2, respectively. The figures show the blocking probability for the two traffic classes as a function of the total normalized offered load. The total load ranges from very small values, close to 0, up to 1, i.e., when the total average bandwidth requested by the two classes is equal to the available capacity.

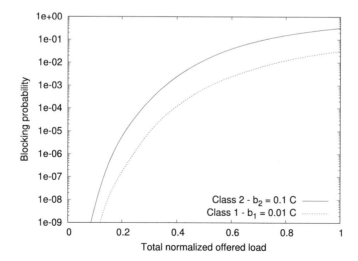

Fig. 4.19 Blocking probability as a function of the total normalized offered load for two classes with bandwidth requirements $b_1 = 0.01\,C$ and $b_2 = 0.1\,C$

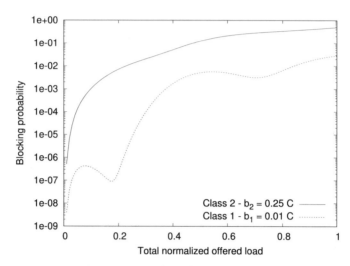

Fig. 4.20 Blocking probability as a function of the total normalized offered load for two classes with bandwidth requirements $b_1 = 0.01\,C$ and $b_2 = 0.25\,C$

The results confirm that the traffic class with lower bandwidth demand always experiences a lower blocking probability, with a mismatch that becomes larger for larger difference in demand. Figure 4.20 also shows that the blocking probability for the class requesting less bandwidth has a peculiar oscillating behavior, resulting in the fact that under some conditions a traffic increase leads to a performance improvement.

This is rather counter-intuitive, as we always expect that an increase in traffic intensity somehow determines a degradation of the performance. The reason of such behavior is a sort of quantization of the resources. When the traffic increases, more calls of traffic classes 1 and 2 arrive. Each class 2 call requires $1/4$ of the total bandwidth C. Now, when the traffic increase is such that it becomes less likely that such an amount of bandwidth is available, the blocking probability for customers of traffic class 2 increases, as expected, while the blocking probability for customers of traffic class 1, which require a much smaller amount of capacity, decreases because they can more easily find some spare capacity that cannot be used to accommodate a class 2 customers request.

Besides having a counter-intuitive effect, the phenomenon described above affects the fairness of the network. Some users may gain a performance advantage with a more aggressive behavior, which is something we would like to avoid. Possible countermeasures that can be put in place are discussed in the next section. For the time being, let us point out that an unbalanced blocking probability also results in an unbalanced network utilization. When the normalized traffic load is the same for both traffic classes, we expect that also the link utilization would be the same. However, this is not the case, as shown in Fig. 4.21: when the offered

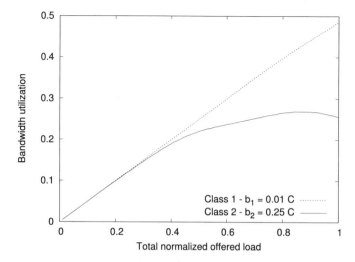

Fig. 4.21 Bandwidth utilization as a function of the total normalized offered load for two classes with bandwidth requirements $b_1 = 0.01\,C$ and $b_2 = 0.25\,C$

load increases, the customers with low bandwidth requirements use the available capacity much more than the customers with high bandwidth requirements.

4.4.3.2   Strategies for QoS Management: Bandwidth Sharing, Bandwidth Partitioning and Trunk Reservation

In the previous case study we found out that different capacity requirements lead to different blocking probability when the calls compete for the available capacity. It is worth asking whether there may be countermeasures to avoid this phenomenon that we can apply to control the blocking probability. A possible objective could be to guarantee better fairness, i.e., all calls should experience the same blocking probability regardless the capacity they request from the network.

The approach adopted in the case analyzed above, where the available bandwidth C is shared among incoming calls without limitations, bounds, or specific scheduling policies, is usually called *complete sharing*. Other policies can be adopted to share the available capacity that can provide tools to control and equalize the blocking probability. For instance, the so-called *complete partitioning* applies an opposite strategy: the bandwidth C is divided into n partitions, and partition C_i is reserved to traffic class i only.

Two other possible strategies are:

- *complete sharing with an ordering constraint*: the capacity C is fully shared, but calls of type i must use less bandwidth than (or the same as) calls of type $i + 1$, i.e.:

$$\mathscr{S} = \{\mathbf{k} : 0 \leq \mathbf{k}\,\mathbf{b} \leq C \quad \text{and} \quad k_i b_i \leq k_{i+1} b_{i+1} \;\; \forall i\}$$

- *partial sharing*: the capacity C is divided into $n + 1$ partitions, partition C_0 is fully shared among all traffic classes while partition C_i is reserved to traffic class i only, i.e.:

$$\mathscr{S} = \left\{ \mathbf{k} : 0 \leq \mathbf{k}\,\mathbf{b} \leq \sum_{i=1}^{n} C_i + C_0 = C \quad \text{and} \quad k_i b_i \leq C_i + C_0 \;\; \forall i \right\}$$

Since the sharing policies mentioned above modify the constraints on the state space, they obviously result in different values of the blocking probability. Unfortunately, given a fixed available capacity C engineering the bandwidth sharing strategy is not an easy problem to solve, since the blocking probabilities of the different classes are still all linked together, and we lack analytical tools that allow to predict the blocking probability of traffic class i as a function of the blocking probability of the other traffic classes. As a matter of fact, the only possible way to proceed would be by iteratively trying and search for a combination of the bounds that provide the desired result. This may end up in a very time consuming task, unpractical when the number of traffic classes is large.

Moreover, the result would be strictly dependent on traffic values, and therefore the effect of the chosen policy would be influenced by the specific traffic load of each class. An example of the effects of the partial sharing strategy is shown in Figs. 4.22, 4.23, 4.24, and 4.25, which show the blocking probability and the bandwidth utilization for the two cases discussed in the previous section. The partial sharing of the link bandwidth is applied assuming the following capacity

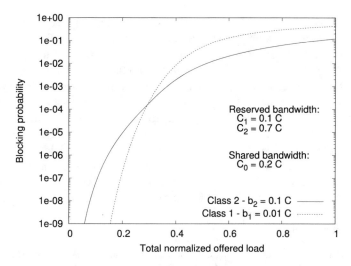

Fig. 4.22 Blocking probability as a function of the total normalized offered load for two classes with bandwidth requirements $b_1 = 0.01\,C$ and $b_2 = 0.1\,C$, assuming a partial sharing strategy with $C_0 = 0.2\,C$, $C_1 = 0.1\,C$, and $C_2 = 0.7\,C$

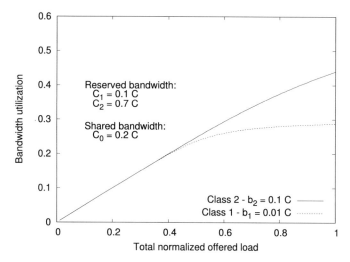

Fig. 4.23 Bandwidth utilization as a function of the total normalized offered load for two classes with bandwidth requirements $b_1 = 0.01\,C$ and $b_2 = 0.1\,C$, assuming a partial sharing strategy with $C_0 = 0.2\,C$, $C_1 = 0.1\,C$, and $C_2 = 0.7\,C$

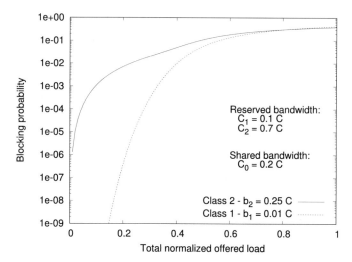

Fig. 4.24 Blocking probability as a function of the total normalized offered load for two classes with bandwidth requirements $b_1 = 0.01\,C$ and $b_2 = 0.25\,C$, assuming a partial sharing strategy with $C_0 = 0.2\,C$, $C_1 = 0.1\,C$ and $C_2 = 0.7\,C$

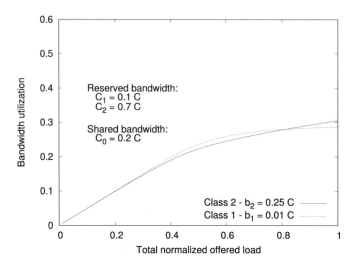

Fig. 4.25 Bandwidth utilization as a function of the total normalized offered load for two classes with bandwidth requirements $b_1 = 0.01\,C$ and $b_2 = 0.25\,C$, assuming a partial sharing strategy with $C_0 = 0.2\,C$, $C_1 = 0.1\,C$ and $C_2 = 0.7\,C$

partitions: $C_0 = 0.2\,C$, $C_1 = 0.1\,C$, and $C_2 = 0.7\,C$. It is apparent that the blocking probability as well as the link utilization becomes significantly unbalanced between the two classes under changing traffic conditions. In particular, with the chosen configuration of the partial sharing strategy the link utilization is equalized in case 2 (Fig. 4.25) but not in case 1 when the load is above 0.5 (Fig. 4.23), and the blocking probability in both cases is equalized only for some specific load regimes (Figs. 4.22 and 4.24).

A different bandwidth sharing policy that proves very effective for equalizing the blocking probabilities and that is also easy to implement is the so-called *trunk reservation*. It works as follows:

- a parameter β_i is chosen as a bandwidth threshold for each traffic class;
- a call of class i is accepted if and only if the residual link capacity C_r is such that $C_r \geq C - \beta_i$, otherwise the call is blocked.

The blocking probabilities π_{pi} can then be equalized by choosing the same value of β_i for all traffic classes. A very common choice is to set the threshold to the maximum used capacity that would allow to accept the class with the highest bandwidth requirement, i.e.:

$$\beta_i = \beta = C - \max_{1 \leq i \leq n} \{b_i\} \quad \forall i$$

The motivation behind this choice can be understood by looking back at Fig. 4.20. If we imagine to split the available capacity C in partitions of size $b_1 = 0.01\,C$, the blocking probability of traffic class 2 is much higher because to accept a call

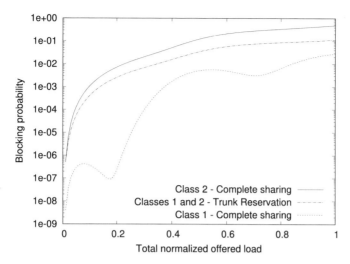

Fig. 4.26 Blocking probability as a function of the total normalized offered load for two classes with bandwidth requirements $b_1 = 0.01\,C$ and $b_2 = 0.25\,C$. Comparison between complete sharing and trunk reservation strategies

at least 25 free bandwidth partition units are needed, whereas to accept a class 1 call only one unit is sufficient. If only 24 partition units are available, a call of class 2 is blocked while 24 calls of class 1 can still be accepted. By setting $\beta = C - \max\{0.01\,C, 0.25\,C\} = 0.75\,C$ also class 1 calls can be accepted if and only if at least 25 partition units are available, as for calls of type 2. The effect of such an approach is to leave some spare capacity by blocking more class 1 calls when the residual capacity is small, which results in having more often at least 25 partition units available for calls of either class 1 or 2. The final result is that π_{p1} increases and π_{p2} decreases, reaching an equal and fair level of call blocking.

Figures 4.26 and 4.27 show the blocking probability and the bandwidth utilization of the two classes with $b_1 = 0.01\,C$ and $b_2 = 0.25\,C$ when trunk reservation is adopted, compared to the values obtained when complete bandwidth sharing is applied. As expected, full fairness is guaranteed between the two classes in terms of blocking and throughput.

4.4.3.3 A Simple Circuit Switching Network

Let us consider the very simple topology illustrated in Fig. 4.28, with $N = 3$ nodes, $J = 2$ trunk groups, and $R = 2$ routes that share trunk group 1. This is a trivial star topology, but it allows for simple computation of the solution. The offered traffic on the two routes is $A_{01} = 0.4$ E and $A_{02} = 0.6$ E, respectively, whereas the trunk group capacities are fixed to $C_1 = 4$ circuits and $C_2 = 3$ circuits. Let us assume that

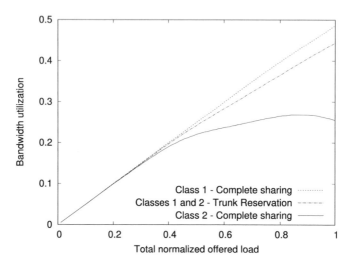

Fig. 4.27 Bandwidth utilization as a function of the total normalized offered load for two classes with bandwidth requirements $b_1 = 0.01\,C$ and $b_2 = 0.25\,C$. Comparison between complete sharing and trunk reservation strategies

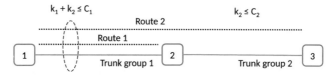

Fig. 4.28 A simple star network topology with $N = 3$ nodes, $J = 2$ trunk groups, and $R = 2$ routes

the calls in this example are simple telephone calls that use one circuit in each trunk group traversed by their route.

The equations that bound the state space are:

$$k_1 + k_2 \leq 4$$
$$k_2 \leq 3$$

We can apply the product form in Eq. (4.96) assuming that the state space is unbounded, obtaining the results presented in Table 4.3, which are limited to the range of interest $k_1, k_2 = 0, 1, 2, 3, 4$.

Table 4.4 reports the product form values in Table 4.3 according to the possible combinations of k_1 and k_2 that determine the state space $\mathscr{S}$. The values corresponding to states in the set $\mathscr{S}_{b1}$ that may lead to a loss of calls on route 1 are highlighted in bold, whereas the values corresponding to states in the set $\mathscr{S}_{b2}$ that may lead to a loss of calls on route 2 are underlined.

Table 4.3 Values of the product forms for an unbounded system with k_1 and k_2 varying from 1 to 4

k_1	k_2 0	1	2	3	4
0	1.0000E+00	6.0000E−01	1.8000E−01	3.6000E−02	5.4000E−03
1	4.0000E−01	2.4000E−01	7.2000E−02	1.4400E−02	2.1600E−03
2	8.0000E−02	4.8000E−02	1.4400E−02	2.8800E−03	4.3200E−04
3	1.0667E−02	6.4000E−03	1.9200E−03	3.8400E−04	5.7600E−05
4	1.0667E−03	6.4000E−04	1.9200E−04	3.8400E−05	5.7600E−06

Table 4.4 Values of the product forms for a bounded state space with $k_1 + k_2 \leq 4$ and $k_2 \leq 3$

k_1	k_2 0	1	2	3
0	1.0000E+00	6.0000E−01	1.8000E−01	3.6000E−02
1	4.0000E−01	2.4000E−01	7.2000E-02	**1.4400E − 02**
2	8.0000E−02	4.8000E−02	**1.4400E − 02**	
3	1.0667E−02	**6.4000E − 03**		
4	**1.0667E − 03**			

We can then calculate:

$$G(\mathscr{S}) = 2.703$$

$$G(\overline{\mathscr{S}_{b1}}) = 2.667$$

$$G(\overline{\mathscr{S}_{b2}}) = 2.631$$

and finally

$$\pi_{p1} = 1.34 \cdot 10^{-2}$$

$$\pi_{p2} = 2.67 \cdot 10^{-2}$$

As expected, the blocking probability is higher for calls that belong to route 2, which crosses two trunk groups including the one with less circuits, having therefore more chances to be blocked. It is worth making a comparison with the analysis that considers only one trunk at a time. Trunk group 1 with $C_1 = 4$ circuits is loaded with $A_0 = A_{01} + A_{02} = 1$ E, leading to $\mathscr{B}(4, 1) = 1.54 \cdot 10^{-2}$. Trunk group 2 with $C_2 = 3$ circuits is loaded with $A_0 = A_{02} = 0.6$ E, leading to $\mathscr{B}(3, 0.6) = 1.98 \cdot 10^{-2}$. As we can see, the values of the blocking probability are not the same as the ones computed with the mBD process model. Given the rather simple case, they are in the same range but with have different values. In particular, with the Erlang $\mathscr{B}$ we cannot find the blocking probability per route, and can only approximate the

Fig. 4.29 Schema of the
network with star topology
and showing the six routes
among the four network sites

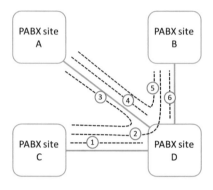

blocking probability per trunk group, which does not tell us what type of call is
actually affected by the blocking.

4.4.3.4   Dimensioning a Small Private Telephone Network

Let us recall the second part of the example in Sect. 4.2.8.3. When a star topology is
adopted to interconnect the PABXs in four sites, we dimensioned the trunk groups
between the nodes independently. That was a rough approximation, and now we
know that the blocking probability has to be calculated per route and not per link or
trunk group. Therefore, let us check how accurate the approximation we made is.

If we consider only the traffic exchanged among the sites, the possible routes are
depicted in Fig. 4.29 and are listed from 1 to 6. The loss probability per route can
be calculated with formula (4.104). This was done by means of a simple computer
program that iteratively calculates the normalizing constants on the different state
spaces. The final results are as follows:

$$G(\overline{\mathscr{S}}_{b1}) = 3.015 \cdot 10^{31}$$
$$G(\overline{\mathscr{S}}_{b2}) = 3.015 \cdot 10^{31}$$
$$G(\overline{\mathscr{S}}_{b3}) = 2.718 \cdot 10^{31}$$
$$G(\overline{\mathscr{S}}_{b4}) = 2.723 \cdot 10^{31}$$
$$G(\overline{\mathscr{S}}_{b5}) = 2.723 \cdot 10^{31}$$
$$G(\overline{\mathscr{S}}_{b6}) = 3.021 \cdot 10^{31}$$

and

$$\pi_{p1} = 1.896 \cdot 10^{-3}$$
$$\pi_{p2} = 1.922 \cdot 10^{-3}$$
$$\pi_{p3} = 1.002 \cdot 10^{-1}$$
$$\pi_{p4} = 9.865 \cdot 10^{-2}$$
$$\pi_{p5} = 9.867 \cdot 10^{-2}$$
$$\pi_{p6} = 2.688 \cdot 10^{-5}$$

As expected, the loss probability per route is not the same for all routes. In particular, routes 1 and 6 are better off because traversing links with less load, as route 2. On the other hand, the routes that traverse the link between sites A and B experience a higher blocking probability being this route more loaded. Indeed we see here the difference of using a route based model against a link based model.

Exercises

1. Consider an ideal system with an infinite number of servers, subject to Poisson arrivals with frequency λ and service times exponentially distributed with average $\bar{\vartheta} = \frac{1}{\mu}$.

   (a) Prove that for $\lambda < \mu$ the most frequent state is always state $k = 0$.
   (b) Assuming that $\lambda = 0.8$ arrivals/s and $\bar{\vartheta} = 1$ s, find the steady state probabilities for states $k = 0, 1, 3, 4$ and the probability of being in any state $k > 4$.
   (c) Assuming that $\lambda = 4$ arrivals/s and $\bar{\vartheta} = 1$ s, find the most frequent state among states $k = 0, 1, \ldots, 8$ and the probability of being in any state $k > 8$.

2. Prove the validity of the recursive Erlang $\mathscr{B}$ formula:

$$\mathscr{B}(0, A_0) = 1$$

$$\mathscr{B}(m, A_0) = \frac{A_0 \, \mathscr{B}(m - 1, A_0)}{m + A_0 \, \mathscr{B}(m - 1, A_0)}$$

3. A circuit switching system without queuing space is equipped with $m = 10$ circuits and subject to Poisson arrivals with frequency $\lambda = 150$ calls/h and service times with average $\bar{\vartheta} = 4$ min. Find:

   (a) the traffic offered to the system;
   (b) the blocking probability;
   (c) the server utilization.

4. Consider a circuit switching node with $N = 6$ input interfaces, $M = 4$ output interfaces and without queuing space. Calls arrive at the input interfaces according to a Poisson process. Each service request keeps one of the output interfaces busy for a given amount of time. The traffic offered to each input interface is $A_0 = 0.2$ E. The node can work in two different ways:

   (a) as a switch, meaning that each request at the input is sent to a specific output, according to some characteristics of the request;
   (b) as a multiplexer, meaning that the pool of M outputs can serve any incoming request.

Assume that in the switch case the probability that a request incoming on input i must be served by output j is uniformly distributed and equal to $\frac{1}{M}$. Compare the blocking probability in the two different operating modes.

5. The internal telephone network of a company distributed across four sites, called A, B, C, and D, must be engineered. A local PABX is installed in each site to switch both intra-site calls and calls exchanged with other sites. The four PABXs must be interconnected with dedicated lines to carry inter-site calls. The estimated traffic among the sites is given in Erlang in the following traffic matrix:

	Source			
Destination	A	B	C	D
A		10	5	8
B	10		5	2
C	5	5		2
D	8	2	2	

The target performance of the network is to guarantee that the call blocking probability is $\pi_p \le 0.01$. Find the minimum number of lines m_{XY} in the trunk group between sites X and Y under the following assumptions:

(a) the network is implemented with a full mesh topology;
(b) the network is implemented with a star topology, where one node reroutes the calls between other pair of nodes.

Compare the results obtained with the two alternatives, finding the most convenient node that should serve as the center of the star topology, and neglecting correlations between the trunk groups when calculating the loss probability.

6. A Wavelength Division Multiplexing (WDM) optical switch has N input and N output fiber interfaces, each providing M wavelength channels supporting lightpath transmission through the network. When a lightpath needs to be established between two end points (e.g., two data centers), each optical switch along the lightpath associates an input channel on the input interface to an output channel on the output interface. The output interface is chosen according to the address of the destination, which is assumed to be uniformly distributed among the output interfaces. The wavelength in the input channel is assigned by the upstream node.
Two different operating modes can be considered:

(a) wavelength continuity: the wavelength of a lightpath cannot change when traversing the switch;
(b) no wavelength continuity: the wavelength of a lightpath can change when traversing the switch using a suitable wavelength conversion technology.

Find the blocking probability when a new lightpath is established in the two operating conditions and compare the results obtained, assuming that the traffic offered by each lightpath is $A_0 = 0.2$ E and that the optical switch size is such that $N = 10$ and $M = 4$.

7. At the beginning of year 2022, a company located in two different sites A and B must be equipped with a telephone network. The expected traffic exchanged between the two sites, as well as the expected traffic exchanged by each site with the external public network, is given in Erlang in the traffic matrix below:

Destination	Source		
	A	B	Ext.
A		4	6
B	4		4
External	6	4	

Based on business growth forecasts, it is expected that the traffic reported above increases by 25% per year. The two sites must be interconnected by dedicated telephone lines to carry inter-site traffic, whereas calls exchanged with the external network must be carried by public lines.

(a) Plan the minimum number of private and public lines to be installed at the beginning of each year until the end of year 2025, such that the call blocking probability is guaranteed to be less than 1% throughout the whole planned period.

(b) Find the utilization of the private lines between A and B at the end of each year considered in the planned period.

8. Two switching systems in the telephone network connect two urban areas that are shifting from a mostly industrial presence to a mostly residential population. This evolution is changing the traffic patterns of the telephone traffic exchanged between the two switching systems. The network operator must plan the number of circuits between the two systems according to the expected growth of the offered traffic, taking into account the worst-case scenario between 10am, considered the peak hour for industrial activities, and 7pm, considered the peak hour for residential customers.

The corresponding average frequencies of call requests, which can be assumed distributed according to a Poisson process, are reported in the following table for each semester in the next 3 years. The calls are assumed to last on average $\bar{\vartheta} = 2\,\text{min}$.

t (months)	λ at 10 am (calls/min)	λ at 5 pm (calls/min)
0	5.00	2.00
6	5.75	2.80
12	6.61	3.92
18	7.60	5.49
24	8.75	7.68
30	10.06	10.76
36	11.57	14.66

Considering the target QoS to guarantee a call blocking probability $\pi_p \leq 0.01$:

(a) find the minimum number m of circuits that must interconnect the two switching systems for the whole 3-year period, assuming that new circuits can be deployed exclusively every six months;

(b) compute the circuit utilization at time $t = 0$ and immediately before any new circuit deployment.

9. A company operates in two sites, called A and B. Each site is equipped with a local PABX. The total offered traffic between the two sites is $A_0 = 8$ E and the connectivity is granted by a group of m circuits. The target QoS to be enforced is a call blocking probability $\pi_p \leq 2\%$. Find the number of circuits m needed to guarantee the target QoS.

Due to the current growth of the company, the forecast is that the traffic increases by 10% each year. Calling $t_0 = 0$ the current time, be $t_1 = t_0 + 1$ year and $t_2 = t_0 + 2$ years. Find the values of the offered traffic $A_0(t_1)$ and $A_0(t_2)$ and the minimum number of circuits needed at t_1 and t_2 to guarantee the target QoS. Given the working conditions at the end of the second year, compute the circuit utilization under two working scenarios:

(a) the circuits are chosen randomly by the calls and therefore on average they experience the same utilization;

(b) the circuits are chosen orderly, from circuit 1 to circuit m, and the two PABXs are equipped with a call packing function that always keeps the circuits with lowest index busy, so that if a call on circuits i ends when there are calls still active in circuits with index $j > i$, these calls are shifted to lower index circuits.

Assuming that, after proper market analysis, a circuit used for less than 30% is found not worth to be leased and carrying its corresponding traffic using the PSTN is actually less expensive, find how many circuits must be leased to guarantee the required performance.

10. A queuing system is subject to Poisson arrivals with frequency $\lambda = 5$ arrivals/min and service times exponentially distributed with average $\bar{\vartheta} = 2$ min. Find:

   (a) the probability of having at least $m = 12$ customers in the system, assuming an infinite number of servers;

   (b) the blocking probability, assuming that the system is equipped with $m = 12$ servers;

   (c) the queuing probability, assuming that the system is equipped with $m = 12$ servers and an infinite queue;

   (d) the average waiting time for any customer and the average waiting time for customers that are being queued, assuming again that the system is equipped with $m = 12$ servers and an infinite queue.

11. A company has 3 different lines of products and provide supports to customers with 3 dedicated call centers, each equipped with $m = 2$ operators. Presently the call centers do not implement any queuing mechanism for the calls arriving when all operators are all busy. The arrival call process to each call center is Poisson with an average arrival rate $\lambda = 0.2$ calls/min, and the call holding times are exponentially distributed with average $\bar{\vartheta} = 2.5$ min.

   (a) Find the call blocking probability π_p per call center.

   (b) Find the call blocking probability assuming that each operator is instructed to provide information on any kind of product, so that all operators can be grouped into one single call center.

   (c) Compare the operator utilization in the two cases described above.

   (d) Determine whether it is necessary to increase or it is possible to decrease the total number of operators m_{tot} if the target blocking probability is $\pi_p \leq 1\%$.

Assume that at some point the call center is updated to implement a mechanism to queue calls waiting for an operator to become available.

   (e) Find the queuing probability π_r in case the total number of operators is kept equal to the last value m_{tot} determined above.

   (f) Determine whether that number m_{tot} can be decreased if the target queuing probability considered acceptable is $\pi_r \leq 20\%$.

   (g) Find the average waiting time $\bar{\eta}$ and the number of lines l to connect the call center to the PSTN in order to guarantee a blocking probability for the incoming calls $\pi_p \leq 1\%$.

12. A company is deploying a call center with waiting space to provide information and assistance to its customers. The call center is connected to the PSTN with $l = 4$ telephone lines, and in order to provide information the company has trained $m = 3$ operators. The traffic offered to the call center can be considered Poissonian and is estimated to be $A_0 = 2$ E. The duration of the conversations with the operators is exponentially distributed with average $\bar{\vartheta} = 4$ min.

   (a) Find the average waiting time for customers who have to wait, assuming that no calls are lost due to the limited number of lines.

   (b) Find the call blocking probability due to a call arriving when all lines are busy.

(c) Determine the actual value of the traffic offered to the operators, subtracting any traffic lost due to the lines found busy.
(d) Recalculate the average waiting time for customers who have to wait using the actual value of the traffic offered to the operators.
(e) Compare the results obtained above in terms of average waiting time and discuss any differences. In case of non-negligible differences, indicate what procedure should be followed in order to correctly assess the actual performance of the call center.

13. A company needs to plan the investments for the deployment of a call center to provide customer support. It is expected that, during peak traffic hours, the call arrival process has a Poisson distribution with an arrival rate of $\lambda = 10$ calls/min, and that the call holding time is distributed exponentially with average $\bar{\vartheta} = 90$ s. The call center must be dimensioned by determining the number l of telephone lines interconnecting it with the PSTN and the number m of call center operators. The quality of service requirements to be met are as follows:

- the blocking probability of incoming calls due to the limited number of lines must be $\pi_p \leq 0.01$;
- the waiting time, for customers who experience waiting, must be $\epsilon \leq \bar{\vartheta} = 90$ s with a probability higher than 90%.

(a) Find the values l_0 and m_0 to be used at the call center installation to guarantee the quality of service requirements, assuming as a first approximation to dimension m_0 that no calls are lost due to the limited number of lines.
(b) Find the values l_1 and m_1 to be used if, after an initial period of operation, the call center has to handle a 20% increase of the offered traffic, still guaranteeing the same quality of service requirements.
(c) Discuss the validity of the approximation made to obtain m_1, verifying what value would be obtained if the loss caused by the l_1 telephone lines were taken into account.

14. A company needs to set up a call center to provide information and assistance to its customers. The call center will be connected to the PSTN with $l = 16$ lines, while the number of operators m to be employed and trained to provide assistance must be determined. It is expected that the traffic offered by the call center customers will be Poissonian and, at peak hours, it value will be $A_0 = 10$ E, and that the duration of the conversations with the operators will be distributed exponentially with an average of $\bar{\vartheta} = 5$ min. The quality requirement to be met to determine the number of call center operators is that the waiting time, for those customers who experience waiting, must be less than 2 min with a probability of at least 85%.

(a) Find the minimum number of operators m to satisfy the quality requirement above, assuming as a first approximation that no calls are blocked due to the limited number of telephone lines.

(b) Find the call blocking probability due to a call arriving when all lines are busy.

(c) Find the actual value of the traffic offered to the operators, as well as the operator utilization.

(d) Discuss the validity of the approximation made to dimension m, verifying what value would be obtained by taking into account the call blocking due to busy lines.

(e) Specify the maximum number of customers that can be simultaneously waiting in the queue and which states of the queuing model used to describe this system are actually unreachable.

Chapter 5
Engineering Packet-Switched Networks

Abstract This chapter introduces the main teletraffic engineering concepts for packet-switched networks, or more generally for networks with dynamic and stochastic bandwidth allocation to traffic flows. The most peculiar characteristic of such systems is that, in general, they can be studied as single server queuing systems. Therefore, in this chapter we first present some general results that are valid for single server systems. Then we deal with the simplest case, the $M/M/1$ system. The formulas to analyze such system are derived from those already developed in the previous chapter for the case of a generic number m of servers. Then we will consider also the case with finite queuing space, i.e., the $M/M/1/L$ system. In the last section we relax the assumption of memoryless service time and the $M/G/1$ system is analyzed, opening the floor to more complex modeling and to the analysis of non-FIFO scheduling policies, such as those based on a priority mechanism.

Keywords Single server queue · Busy period · Idle time · Server utilization · $M/M/1$ system · Waiting time for generic customers · Waiting time for queued customers · $M/M/1/L$ system · General service time · $M/G/1$ system · Priority scheduling · Pre-emption · Non pre-emption · Shortest job next

5.1 Introduction

The main characteristics of a typical packet switching system are:

- segmentation of the information flow into sub-blocks called packets, segments, datagrams, etc. that carry at once both data and signaling information, similarly to the traditional mailing service;
- dynamic bandwidth sharing based on the instantaneous demand of multiple traffic flows.

There are indeed other cases, for instance, slotted systems with centralized algorithms for bandwidth allocation, i.e., with a round robin scheduling, or packet switching systems where several users share the same channel (the typical wireless LAN case). These cases require specific modeling that are beyond the scope of this

© Springer Nature Switzerland AG 2023
F. Callegati et al., *Traffic Engineering*, Textbooks in Telecommunication Engineering, https://doi.org/10.1007/978-3-031-09589-4_5

book. Here we stick to the case of many users sending packets to a packet switching system which stores and forwards them on output serial links.

Generally speaking, the interfaces to packet switching systems (like our personal computers or smartphones connecting to the Internet) logically see the channel capacity as a serial bus that can be fully used to transmit one packet at a time. The faster the channel, the shorter the packet transmission time. Packets arriving when the channel (i.e., the server) is busy transmitting another packet are queued.

In this chapter we split the study of packet switching systems into four main cases. In Sect. 5.3.1 we present the classical $\mathcal{M}/\mathcal{M}/1$ queuing system, by simply recalling the formulas already developed in the previous chapter for the $\mathcal{M}/\mathcal{M}/m$ system and applying them to the case of $m = 1$. The single server system is a model that, in spite of its simplicity, provides results that lead to understanding the main concepts in packet-switched networks, when applied to simple case studies like the outbound interface of a router or the interconnection link between two LANs. Then in Sect. 5.3.2 we look at the case of limited queue size and present the $\mathcal{M}/\mathcal{M}/1/L$ system. In Sect. 5.4.1 we relax the assumption of memoryless service time, thus introducing a more realistic modeling approach, since almost every packet switching technology has well defined boundaries to the size of the packets. We study the $\mathcal{M}/\mathcal{G}/1$ queuing system and provide guidelines and engineering tools for typical packet switching protocols and use cases. Finally, in Sect. 5.4.2 we study the problem of introducing more complex scheduling policies, providing some case studies taken from modern telecom as well as computer architectures.

5.2   Single Server Queuing

5.2.1   Performance Metrics

Let us briefly discuss which are the performance metrics we are interested in. Generally speaking, we assume that packet switching is implemented according to the store-and-forward paradigm: packets are stored at arrival, their header is analyzed, and the decision on how to forward them is taken accordingly. Then the packet is sent to the output channel according to the proper forwarding decision.

We assume that the store-and-forward device, such as an IP router, is properly dimensioned with enough memory and processing capabilities to match the amount of packets that need to be stored, processed, and forwarded. Therefore, the main focus here is not on the loss of customers (i.e., the packets) but on delay, i.e., how much time a packet has to wait in the router's output queue before being forwarded, and on the probability that a queuing event happens. The delay, often also called *latency*, may not be negligible and may increase significantly when many nodes must be traversed, leading to impairments to the quality of service perceived by the user (either a human or a computer). A problem that is becoming even more critical with the support of more complex services, such as multimedia communications,

IoT sensing systems, communications in critical infrastructures, and other time sensitive application scenarios.

Therefore, considering our initial focus on single server queues with infinite waiting space, the performance metrics of interest are the delay due to the queuing that a customer may experience, and the probability that a customer has to wait. The choice of an infinite queue is important to simplify the analysis since it makes the problem more tractable from the analytical point of view. Moreover, it allows the reuse of many results we have already presented in Chap. 4

5.2.2 A General Result for Single Server Systems with Infinite Waiting Space

A general result holds for any single server systems, which depends only on the characteristic of having $m = 1$ and not on the traffic profile and/or service time distribution. The only requirement is that the service times are "independent and identical distributed" (i.i.d.). Therefore, we now refer to the very general $\mathcal{G}/\mathcal{G}/1$ queue.

If we observe a single server queuing system for a period of time T, we can define:

- **busy time** $B(T)$ the subset of T when the server is busy, i.e., the sum of all the service times occurring during T;
- **idle time** $I(T)$ the subset of T when the server is idle, i.e., the sum of all the time intervals within T in which the server does not have any packet to transmit.

Given that in a queuing system with FIFO scheduling the server can be either idle or busy and not in other states, it obviously holds that:

$$B(T) + I(T) = T$$

Figure 5.1 shows a graphical example of the concepts of busy and idle time. Given that $m = 1$, during T the number of customers in service takes the values 0 or 1 randomly, according to the customer behavior. The different busy periods, each of which may be due to a single service or to several services in a row, altogether sum up and determine the overall busy period $B(T)$. Generally speaking, $B(T)$ and $I(T)$ are random variables, and $\bar{B}(T)$ and $\bar{I}(T)$ are their average values.

At the same time, if the system is ergodic and its behavior can be described by means of steady state probabilities, let P_0 be the probability that the server is idle. It follows that:

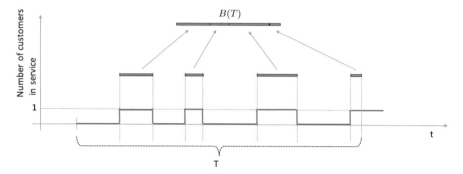

Fig. 5.1 Example of busy and idle time during a generic time period T

$$\bar{B}(T) = T(1 - P_0) \qquad \bar{I}(T) = T P_0$$

If the average service time is $\bar{\vartheta} = \frac{1}{\mu}$, it follows that the average number of users served in T, i.e., the average number of departures during T, is given by:

$$\bar{d}(T) = \frac{\bar{B}(T)}{\bar{\vartheta}} = \frac{T(1 - P_0)}{\bar{\vartheta}} = T(1 - P_0)\mu$$

Since an ergodic system works in a statistical equilibrium and each customer that enters the system must exit from the system once the service is complete, the average number of departures in T must be equal to the average number of arrivals in T, i.e.:

$$\bar{d}(T) = \bar{a}(T)$$

If the arrival process has a finite average arrival rate, it follows that:

$$\bar{a}(T) = \lambda T$$

Therefore:

$$T(1 - P_0)\mu = \lambda T \tag{5.1}$$

from which we find that:

$$P_0 = 1 - \frac{\lambda}{\mu}$$

As usual, the offered traffic is $A_0 = \frac{\lambda}{\mu}$ and, in case of an infinite queue with a single server, $A_s = A_0 = \rho$. Therefore:

$$P_0 = 1 - A_0 = 1 - \rho$$

In practice, for any ergodic single server queuing system the probability of the server to be idle is the same, i.e.:

$$P_0 = 1 - \rho \tag{5.2}$$

The result obtained above requires only that an average arrival rate λ and an average service time $\bar{\vartheta}$ exist, and that the system is ergodic. An alternative way to find (5.2) is to observe that, for an ergodic system where the statistical average is equal to the time average, we can write:

$$\rho = \frac{\bar{B}(T)}{T} = \frac{\bar{B}(T)}{\bar{B}(T) + \bar{I}(T)} = 1 - P_0 \tag{5.3}$$

5.3   Memoryless Single Sever Queuing Systems

5.3.1   Infinite Queuing Space: The $\mathcal{M}/\mathcal{M}/1$ System

The $\mathcal{M}/\mathcal{M}/1$ queuing system is a single server queuing system with Poisson arrivals and exponential service times. We have already studied this case in the previous chapter for a generic value of m, therefore it is relatively easy to apply the related formulas to the case of $m = 1$. Indeed, Eq. (4.49) becomes:

$$\lambda_k = \lambda \;\; \forall k \geq 0 \quad \text{and} \quad \mu_k = \mu \;\; \forall k > 0 \tag{5.4}$$

Moreover:

$$A_0 = \frac{\lambda}{\mu} = A_s = \rho \tag{5.5}$$

From the expressions (3.10) and (3.11) the steady state probabilities are:

$$P_k = P_0 \, A_0^k = P_0 \, \rho^k \quad \forall k \geq 0$$
$$P_0 = \frac{1}{\sum_{k=0}^{+\infty} \rho^k} = 1 - \rho \quad \text{if } \rho < 1 \tag{5.6}$$

Note that $P_0 = 1 - \rho$, as expected from the discussion in the previous section. According to the conventions adopted in this book, we should write the steady state probabilities as $P_k = (1 - A_0) \, A_0^k$. Nonetheless, in the traditional literature about single server queues they are mostly reported as $P_k = (1 - \rho) \, \rho^k$ and for this reason we will follow such notation, as long as $A_0 = \rho$. As expected, P_0 has a finite value if and only if $\rho < 1$, i.e., when $\lambda < \mu$. Under this assumption the system is ergodic and the steady state probabilities exist.

5.3.1.1 Congestion: The Probability of Being Queued

The first performance metric of interest is the probability of a packet being queued. Thanks to the PASTA property, this is the probability that the server is busy, i.e., the probability that $k \geq 1$:

$$\pi_r = \Pr\{k \geq 1\} = \sum_{k=1}^{\infty} P_k = 1 - P_0 = \rho$$

This can also be obtained from the Erlang $\mathscr{C}$ formula: $\mathscr{C}(1, A_0) = \mathscr{C}(1, \rho) = \rho$. The average number of customers in the system (i.e., the traffic in the system) is:

$$A = \sum_{k=0}^{\infty} k\, P_k = \sum_{k=0}^{\infty} k(1 - \rho)\rho^k$$

$$= (1 - \rho)\rho \sum_{k=0}^{\infty} k\, \rho^{k-1} = (1 - \rho)\rho \frac{d}{d\rho} \sum_{k=0}^{\infty} \rho^k$$

$$= (1 - \rho)\rho \frac{d}{d\rho} \frac{1}{1 - \rho} = \frac{\rho}{1 - \rho} \qquad (5.7)$$

The traffic A is plotted in Fig. 5.2 as a function of ρ. As intuition suggests, A goes to infinity as ρ approaches 1, since this system has a maximum throughput equal to 1 (i.e., when the single server works all the time). If $\rho \to 1$ the probability that a customer arrives and finds the server busy is almost 1, all customers are queued and the size of the queue sky-rockets. If ρ reaches 1 or goes beyond that, then the system is not ergodic anymore, the queue grows to infinity and the steady state probabilities do not exist.

5.3.1.2 Delay: The Time Spent Waiting in the Queue

The probability of being queued is a performance metric of interest, but the time spent in the queue waiting to be served is by far more important. It is worth clarifying what exactly we mean when we want to quantify the time spent in the queue. Similarly to what already discussed in Sect. 4.3, there are two quantities of interest:

- the global waiting time in the queue: a random variable η that may take values between 0 and ∞, including 0, and refers to all customers entering the system;
- the waiting time in the queue conditioned to the event of waiting: a random variable ϵ that is always larger than 0, and refers to the subset of customers that do not find the server free and must therefore be queued.

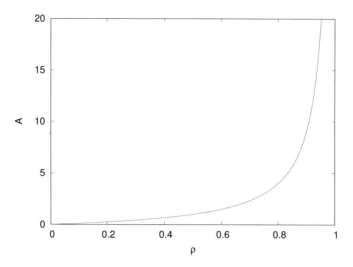

Fig. 5.2 $\mathcal{M}/\mathcal{M}/1$: A as a function of ρ

Obviously, η and ϵ are different metrics and, if we take the average values, we expect that $\bar{\epsilon} \geq \bar{\eta}$. Generally speaking, it is not possible to say whether η is more or less meaningful than ϵ, since this depends on the specific application and QoS goal. Nonetheless, it is important to be aware that, when dealing with waiting time, it is necessary to clarify how we intend to compute it, i.e., whether including or not the customers that do not wait at all.

Now let us start with the average values, and in particular with $\bar{\eta}$. It can be calculated in several ways, which are somehow interesting to explore as an exercise.

At first, let us start from Little's Theorem. The throughput of the system is $A_s = \rho$ and therefore, once we know the traffic in the system A as in Eq. (5.7), the average number of customers in the queue can be obtained as:

$$A_c = A - A_s = \frac{\rho^2}{1 - \rho} \tag{5.8}$$

Since there are no lost customers, $\lambda = \lambda_s$ and then:

$$\bar{\eta} = \frac{A_c}{\lambda} = \frac{\rho}{\lambda} \frac{\rho}{1 - \rho} = \bar{\vartheta} \frac{\rho}{1 - \rho} \tag{5.9}$$

Similarly, we can calculate A_c following the same procedure as in Eq. (5.7), but considering that when there are k customers in the system there are $k - 1$ customers in the queue. Therefore:

$$A_c = \sum_{k=1}^{\infty}(k-1)P_k = (1-\rho)\sum_{k=1}^{\infty}(k-1)\rho^k$$

$$= (1-\rho)\rho^2\sum_{k=1}^{\infty}(k-1)\rho^{k-2} = (1-\rho)\rho^2\frac{d}{d\rho}\sum_{h=0}^{\infty}\rho^h = \frac{\rho^2}{1-\rho} \qquad (5.10)$$

and then we obtain $\bar{\eta}$ as in (5.9).

Note that to calculate formulas (5.8) and (5.10) no assumptions were made regarding the queue scheduling policy, therefore the results are valid for any scheduling policy. If we assume FIFO scheduling, we can obtain the same result looking at different values of the delay. In a FIFO queue, if a customer enters the system and finds k customers already there, thanks to the memoryless property of the exponential service time the newly arrived customer stays in the queue on average for an amount of time given by $k\bar{\vartheta}$. Therefore, we can say that the average waiting time conditioned to the number of customers found in the system is:

$$\bar{\eta}_{|k} = k\bar{\vartheta}$$

and then we can write:

$$\bar{\eta} = \sum_{k=0}^{\infty}\bar{\eta}_{|k}P_k = \sum_{k=0}^{\infty}k\bar{\vartheta}(1-\rho)\rho^k = \bar{\vartheta}(1-\rho)\rho\sum_{k=0}^{\infty}k\rho^{k-1} = \bar{\vartheta}\frac{\rho}{1-\rho} \qquad (5.11)$$

that obviously gives us the same result as in Eq. (5.9).

In Fig. 5.3 $\bar{\eta}$ is plotted as a function of ρ varying $\bar{\vartheta}$ as a parameter. Indeed, if $\bar{\vartheta} = 1$ the curve is identical to that of Fig. 5.2 and can be used as a sort of benchmark. If $\bar{\vartheta} \neq 1$, then $\bar{\eta}$ will be larger or smaller than this benchmark. It is rather obvious from Eq. (5.9) that, for a given value of the load ρ, the average waiting time $\bar{\eta}$ is proportional to the average service time $\bar{\vartheta}$, as also shown in Fig. 5.4. Figures 5.3 and 5.4 show an important property of this queuing system, that is a general concept valid for similar systems as well: given a value of load ρ, then the actual value of $\bar{\eta}$ is proportional to the value of $\bar{\vartheta}$, therefore *the average waiting time is shorter when the average service time is shorter*.

Let us now consider the waiting time limited to the customers that have to wait. To calculate it, let us first find the steady state probabilities conditioned to the event of being queued. Indeed the state space is the same as the previous one, except for state $k = 0$ which is the only state that does not lead to queuing. Therefore, the state space is *truncated*, similarly to what happened for the multi-dimensional BD process discussed in Sect. 4.4. Therefore, the only difference is in the normalizing constant and the conditioned steady state probabilities turn out as:

$$P_{k|k\geq 1} = \frac{1-\rho}{\rho}\rho^k$$

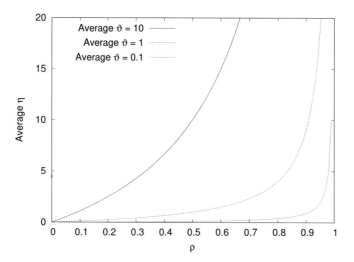

Fig. 5.3 $\mathcal{M}/\mathcal{M}/1$: $\bar{\eta}$ as a function of ρ varying $\bar{\vartheta}$ as a parameter

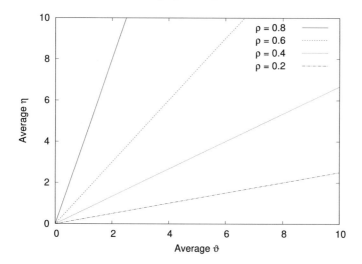

Fig. 5.4 $\mathcal{M}/\mathcal{M}/1$: $\bar{\eta}$ as a function of $\bar{\vartheta}$ varying ρ as a parameter

It follows that:

$$A_{c|k\geq 1} = \sum_{k=1}^{\infty}(k-1)P_{k|k\geq 1} = \frac{1-\rho}{\rho}\sum_{k=1}^{\infty}(k-1)\rho^k = \frac{\rho}{1-\rho} \qquad (5.12)$$

and then:

$$\bar{\epsilon} = \frac{A_{c|k\geq 1}}{\lambda} = \frac{\bar{\vartheta}}{1-\rho} \qquad (5.13)$$

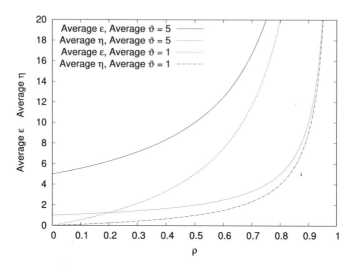

Fig. 5.5 $\mathcal{M}/\mathcal{M}/1$: $\bar{\eta}$ compared to $\bar{\epsilon}$ as a function of ρ, for $\bar{\vartheta} = 5$ and $\bar{\vartheta} = 1$

As expected:

$$\bar{\epsilon} = \frac{\bar{\eta}}{\rho} \geq \bar{\eta}$$

In particular, the difference between $\bar{\epsilon}$ and $\bar{\eta}$ is larger for smaller values of ρ, which makes sense since the smaller the load, the smaller the percentage of queued customers. This means that there are more events contributing to $\bar{\eta}$ with $\eta = 0$ that are not taken into account in the calculation of $\bar{\epsilon}$. The different trends of $\bar{\eta}$ and $\bar{\epsilon}$ as a function of ρ for two values of $\bar{\vartheta}$ is shown in Fig. 5.5.

Finally, recalling formulas (4.79), (4.81) and (4.83), it is easy to obtain the probability density function and the probability distribution of η, ε e δ:

$$f_\eta(t) = (1 - \rho)\delta(t) + \rho(\mu - \lambda)e^{-(\mu-\lambda)t} \tag{5.14}$$

$$f_\varepsilon(t) = (\mu - \lambda)e^{-(\mu-\lambda)t} \tag{5.15}$$

$$f_\delta(t) = (\mu - \lambda)e^{-(\mu-\lambda)t} \tag{5.16}$$

Note that ε and δ are two different quantities but, in this particular case, they turn out to have exactly the same value and probability distribution.

5.3.1.3   The Output Traffic Process: Burke's Theorem

Up to now we have always discussed arrival processes and we never focused on the *departure process*, i.e., the random process that describes the departures from the system. For the $\mathcal{M}/\mathcal{M}/1$ queue there is an interesting result, that we are not going to prove formally but simply describe and discuss because it is of some interest to gain more insight on the topic.

It can be proven that, when considering the *time instant of the departures*, this is a *random process that corresponds to a Poisson process with rate* λ, therefore identical from the statistical point of view to the arrival process.

This result is known as *Burke's theorem*[1] and at first it may look somehow counter-intuitive. Given that μ is the service rate, a first glimpse to the issue could lead to the wrong conclusion that the departure process must have a rate that is also μ. However, this is not the case: μ is the service rate, i.e., the departure rate when there are customers to serve. Due to the random nature of arrivals, sometimes the server is idle and the service rate drops to 0. Therefore, on average we have a departure rate equal to μ with probability ρ (that is the probability of the server being busy), and a departure rate that is 0 with probability $1 - \rho$. If we average these two values, it turns out that the average departure rate is:

$$\mu \cdot \rho + 0 \cdot (1 - \rho) = \lambda$$

which is in accordance to Burke's Theorem in terms of average values.

It is more difficult to prove that the process is also Poissonian in nature. Nonetheless, it is important to note that the incoming and outgoing traffic flows are not exactly the same, which is also a chance to better understand the nature of the traffic considered up to now. As a matter of fact, as explained in Sect. 2.4, for the incoming traffic profile considered so far the service time is completely independent of the inter-arrival rate. When we look at a traffic that, passing through a single server queue, is departing from it, this is not the case anymore. Packets can be distributed in time according to a Poisson process, as stated by Burke's Theorem, but service times are not drawn randomly anymore. As a matter of fact, the service time of a packet in the outgoing traffic profile *must* be smaller than the inter-arrival time between the previous departure and the current one. The reason is that the server can serve a customer (packet) at a time. Therefore, a new packet cannot leave the system before the service of the previous one is completed.

Figure 5.6 graphically explains this concept. Some random arrivals at a single server queue are represented in the top line. The arrows outline the arrival instants and the bars on top of them provide a graphical representation of the service times. Departure times are instead plotted in the bottom line, again as arrows. Indeed, the service time of the different customers in the bottom line must be represented in

---

[1] From P. J. Burke, who was the first to demonstrate it in 1956.

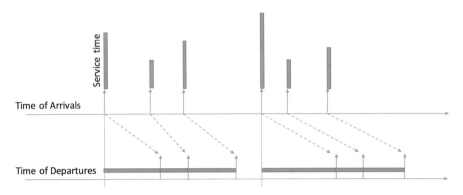

Fig. 5.6 Graphical example of the input and output traffic from a single server queue

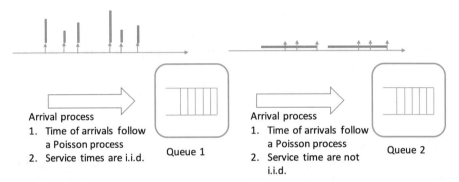

Fig. 5.7 Two queues in sequence: the arrival process to the second queue shows a correlation between inter-arrival times and service times that does not cause any queuing

sequence, since they cannot overlap because of the single server available. This creates a correlation between departure instants and service times of the customers.

This is an important issue in case we are interested in studying multiple queues at once. If we imagine a cascade of two or more queuing systems in sequence, as in the example of Fig. 5.7, the output traffic of one system feeds the next one. Because of the correlation existing between departure times and service times, the input traffic to the second queue arrives according to a Poisson process, if we focus on the arrival time only, but the service times are not independent and no further arrival occurs during the service time of each packet. As a consequence, packets are never queued in the second queuing system.

5.3.2 Finite Queuing Space, the $\mathcal{M}/\mathcal{M}/1/L$ System

The purpose of this section is to understand what happens to the $\mathcal{M}/\mathcal{M}/1$ queue when the queuing space is not infinite, but limited to a maximum of L customers. The steady state probabilities of the $\mathcal{M}/\mathcal{M}/1/L$ system can still be obtained using formula (3.11), limiting it to the $L + 2$ existing states and using a new normalizing constant P_0. Therefore:

$$P_k = P_0 \left(\frac{\lambda}{\mu}\right)^k = P_0 A_0^k \quad 0 \le k \le L + 1 \tag{5.17}$$

and

$$P_0 = \frac{1}{\sum_{k=0}^{L+1} A_0^k} = \begin{cases} \dfrac{1 - A_0}{1 - A_0^{L+2}} & \text{when } A_0 \neq 1 \\ \dfrac{1}{L + 2} & \text{when } A_0 = 1 \end{cases} \tag{5.18}$$

Due to the finite number of states, there are no stability issues and the steady state probabilities exist for any value of A_0.

5.3.2.1 Blocking Probability

Given the limited number of places in the queuing system, there is a certain percentage of customers that are not able to enter in the system. Therefore, the loss probability π_p is a performance metric of major importance for this system. As usual, thanks to the PASTA property, the packet loss probability is given by the probability that the system is full, i.e.:

$$\pi_p = P_{L+1} = \begin{cases} \dfrac{1 - A_0}{1 - A_0^{L+2}} A_0^{L+1} & \text{when } A_0 \neq 1 \\ \dfrac{1}{L + 2} & \text{when } A_0 = 1 \end{cases} \tag{5.19}$$

Figures 5.8 and 5.9 show the blocking probability π_p as a function of L and A_0, respectively. Figure 5.8 is plotted in semi-logarithmic scale and shows that, when $A_0 < 1$ the blocking probability decreases exponentially when L increases, meaning that an arbitrary small value of π_p can be achieved as long as a queuing space large enough is put in place. This is not true anymore for $A_0 \ge 1$, when π_p shows an asymptotic behavior for $L \to \infty$. To understand this phenomenon, let us note that when $A_0 > 1$ we can write:

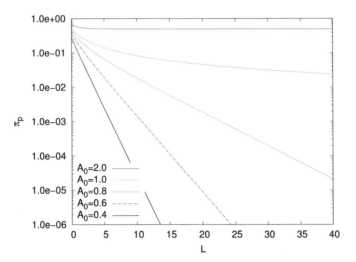

Fig. 5.8 $\mathcal{M}/\mathcal{M}/1/L$: π_p as a function of L varying A_0 as a parameter

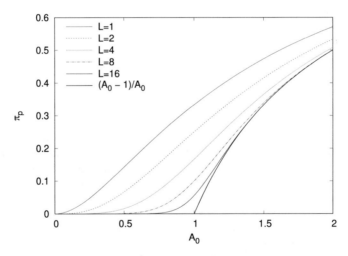

Fig. 5.9 $\mathcal{M}/\mathcal{M}/1/L$: π_p as a function of A_0 varying L as a parameter, including the asymptote for $L \to \infty$

$$\lim_{L\to\infty} \pi_p = (1-A_0) \lim_{L\to\infty} \frac{A_0^{L+1}}{1 - A_0^{L+2}} = (1-A_0) \lim_{L\to\infty} \frac{A_0^{L+1}}{A_0^{L+1}} \frac{1}{\frac{1}{A_0^{L+1}} - A_0} = \frac{A_0 - 1}{A_0}$$

This means that when the queue is overloaded, i.e., for $A_0 > 1$, the blocking probability cannot be controlled, regardless of the size of the queue. Since the system has only one server, the maximum throughput is $A_s^{\max} = 1$. In general the lost traffic is $A_p = A_0 - A_s$ and, in the specific case considered here, it can be at least $A_p^{\min} = A_0 - A_s^{\max}$. Therefore, the blocking probability shows a lower bound at $\pi_p^{\min} = \frac{A_p^{\min}}{A_0}$.

Finally, it is worth highlighting that, when $A_0 \to \infty$ the system ends up being always full, and then:

$$\lim_{A_0 \to \infty} P_k = \begin{cases} 0 & \text{when } k < L+1 \\ 1 & \text{when } k = L+1 \end{cases} \tag{5.20}$$

Note also that when $L = 0$, then $\pi_p = \frac{A_0}{1+A_0}$, which obviously coincides with the result obtained applying the Erlang $\mathscr{B}$ formula $\mathscr{B}(1, A_0)$.

5.3.2.2 Average Performance Metrics

The average server utilization (or average number of customers in service) is:

$$\rho = A_s = A_0(1 - \pi_p) = A_0 \left(1 - \frac{1 - A_0}{1 - A_0^{L+2}} A_0^{L+1} \right) = A_0 \frac{1 - A_0^{L+1}}{1 - A_0^{L+2}} = 1 - P_0 \tag{5.21}$$

whereas the average number of customers in the system (the traffic A) is:

$$A = \sum_{k=1}^{L+1} k P_k = \sum_{k=1}^{L+1} P_k + \sum_{k=1}^{L+1} (k-1) P_k = 1 - P_0 + P_0 A_0^2 \sum_{k=1}^{L+1} (k-1) A_0^{k-2}$$

$$= 1 - P_0 + P_0 A_0^2 \frac{d}{dA_0} \sum_{k=1}^{L+1} A_0^{k-1} = 1 - P_0 + P_0 A_0^2 \frac{d}{dA_0} \frac{1 - A_0^{L+1}}{1 - A_0}$$

$$= 1 - P_0 + P_0 A_0^2 \frac{-(L+1) A_0^L (1 - A_0) + 1 - A_0^{L+1}}{(1 - A_0)^2}$$

$$= \frac{A_0}{1 - A_0} \frac{1 - (L+2) A_0^{L+1} + (L+1) A_0^{L+2}}{1 - A_0^{L+2}} \tag{5.22}$$

The traffic A as a function of A_0 is shown in Fig. 5.10 varying L as a parameter, and it is compared to the case of an $\mathscr{M}/\mathscr{M}/1$ system where $L = \infty$. As expected, A shows the largest values for an infinite queue and decreases for smaller queue sizes. In case of a finite queue size, when A_0 reaches 1 then A does not go to infinity but tends to $L + 1$ when $A_0 \to \infty$, i.e., to a system completely full.

Finally, combining (5.21) and (5.22) we obtain:

$$A = A_s + P_0 A_0^2 \frac{-(L+1) A_0^L (1 - A_0) + 1 - A_0^{L+1}}{(1 - A_0)^2}$$

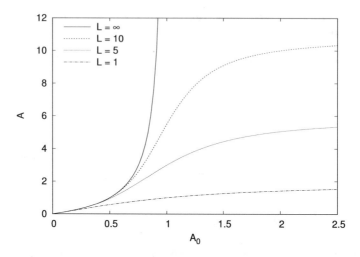

Fig. 5.10 $\mathcal{M}/\mathcal{M}/1/L$: the traffic A as a function of A_0 varying L as a parameter

and since $A = A_s + A_c$ we get:

$$A_c = P_0\, A_0^2 \frac{-(L+1)\, A_0^L\, (1-A_0) + 1 - A_0^{L+1}}{(1-A_0)^2} \tag{5.23}$$

5.3.3   Examples and Case Studies

5.3.3.1   LAN Interconnection with a VPN

The goal is to build an extended Local Area Network (LAN) between two remote sites A and B, to be connected by means of a Virtual Private Network (VPN) service with a guaranteed bit rate. Each site is equipped with a LAN and a VPN router connected to the other site by means of a geographical link with guaranteed capacity $C = C_0 = 100$ Mbit/s. The channel capacity of the LANs is $C_L \gg C_0$, therefore when packets arriving to a router must enter a LAN, the input bit rate is much lower than the output bit rate and we can assume there is no queuing. On the other hand, when packets arriving to a router must leave a LAN, the input bit rate is larger than the output bit rate, therefore queuing may occur.

Let us assume that the inter-LAN traffic is symmetric and packets from A to B (and from B to A) arrive according to a Poisson process with average arrival rate $\lambda = 5000$ packets/s. As a first step approximation, let us assume that packets have a random length that is exponentially distributed with average $D = 800$ bytes.

At first we want to find the average waiting time for the packets that have to wait at the VPN router output interface.

Then let us assume that during the next two years the inter-LAN traffic will grow at the pace of 40% per year in terms of packet arrival rate. We need to decide whether C must be increased at $t = 0$ or $t = 1$ year to keep the performance unchanged, in spite of the traffic increase. In case this is necessary, we must consider that the VPN guaranteed bit rate can be acquired only as a multiple of C_0.

First of all, let us compute the average service time:

$$\bar{\vartheta} = \frac{D}{C_0} = \frac{6400}{100 \cdot 10^6} = 64 \ \mu s$$

and therefore:

$$\rho = 5 \cdot 10^3 \cdot 64 \cdot 10^{-6} = 0.32$$

Then we can obtain the average waiting time for the packets that have to wait:

$$\bar{\epsilon} = \frac{\bar{\vartheta}}{1 - \rho} = 94 \ \mu s$$

Now let us calculate the new values of λ when the traffic increases:

$$\lambda(1) = 5000 \cdot 1.4 = 7000 \text{ packets/s}$$

$$\lambda(2) = 7000 \cdot 1.4 = 9800 \text{ packets/s}$$

Finally, let us check what is the guaranteed bit rate of the VPN connection that will keep $\bar{\epsilon} \leq \epsilon_0 = 100 \ \mu s$:

$$\frac{\frac{D}{C}}{1 - \lambda \frac{D}{C}} \leq \epsilon_0$$

resulting in:

$$C \geq \frac{D}{\epsilon_0}(1 + \lambda \epsilon_0)$$

so that:

$$C(1) = 108.8 \text{ Mbit/s}$$

$$C(2) = 126.7 \text{ Mbit/s}$$

As a matter of fact, the VPN connection must be dimensioned with $C = 2C_0$ already at $t = 0$, since C_0 will not satisfy the performance constraint all over the first year, i.e., up to time $t = 1$. On the other hand, a capacity $C = 2C_0$ will guarantee the requested performance all over the time span of the two years.

5.3.3.2   IoT Data Collection

Let us consider an IoT scenario where subsets of remote sensors send data to a local concentrator (or gateway) T, which will then communicate with a monitor system R. Let us assume that there are $N = 5$ concentrators collecting data from an equal number of sensors in similar environments, and that each concentrator has its dedicated radio channel toward the monitor system. Let us imagine that this is a simple monitoring system without feedback, therefore we are concerned only with the $T \rightarrow R$ communication direction. At the T side all messages from the sensors are stored locally, processed for data conditioning and then sent to R, where all arriving packets are stored in a single queue and processed by a common processing unit according to a FIFO scheduling policy. In both T and R systems, the memory used to store the packets can be assumed infinite in size. The overall IoT data collection system can then be represented as in Fig. 5.11.

Let us assume that packets have a random size that is exponentially distributed with an average $D = 512$ bits, and that the overall packet arrival process from all sensors can be assumed as a Poisson process with rate $\lambda = 100$ packets/sec. Our objectives are as follows:

1. Dimension the bit rate C_T of the radio channels between the concentrators and the monitor in such a way that the average time spent in the concentrator by the packets is $\bar{\delta}_T \leq 0.1$ s. The speed C_T has to be chosen in the following set of normalized speeds: $(2400, 4800, 9600, 14,400, 19,200)$ bit/s.

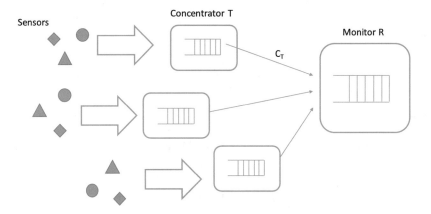

Fig. 5.11 Graphical example of the system considered in the IoT data collection use case

2. Dimension the processing speed μ_R at the monitor system R such that the average time spent in the whole data collection system is $\bar{\delta}_S \leq 0.2$ s, including the average time spent in the concentrator and the average time spent in the monitor.
3. With the values determined above, assume a finite queue length in R and find its minimum size L that can guarantee a packet loss probability $\pi_p \leq 10^{-6}$.

Being all equal, each radio channel between a concentrator and the monitor can be studied as an isolated $\mathcal{M}/\mathcal{M}/1$ system. Since the packets are randomly and uniformly sent by the sensors to the N concentrators, each $\mathcal{M}/\mathcal{M}/1$ system experiences a packet arrival rate $\lambda_T = \frac{\lambda}{N}$.

The service rate at T is $\mu_T = \frac{C_T}{D}$ and then:

$$\bar{\delta}_T = \frac{\bar{\vartheta}_T}{1 - \rho_T} = \frac{1}{\mu_T - \lambda_T} = \frac{1}{\frac{C_T}{D} - \frac{\lambda}{N}} \leq 0.1\text{s}$$

and we find:

$$C_T \geq \left(\frac{\lambda}{N} + 10\right) D = 15360 \text{ bit/s} \tag{5.24}$$

leading to the choice of:

$$C_T = 19200 \text{ bit/s}$$

resulting in:

$$\bar{\delta}_T = 0.057 \text{ s}$$

For the monitor R we can use the Burke theorem (see Sect. 5.3.1.3) to say that the overall process is still a Poisson one in terms of arrivals. Moreover, the mix of packets from the N concentrators makes the service times almost independent of the overall arrival process. Therefore, the total arrival rate is λ and R can be studied as an $\mathcal{M}/\mathcal{M}/1$ system as well.

We can obtain the total time spent in the system as:

$$\bar{\delta}_S = \bar{\delta}_T + \bar{\delta}_R = \frac{1}{\mu_T - \lambda_T} + \frac{1}{\mu_R - \lambda_R}$$

where μ_T is known already, and therefore:

$$\frac{1}{\frac{C_T}{D} - \frac{\lambda}{N}} + \frac{1}{\mu_R - \lambda} \leq 0.2 \text{ s}$$

that gives:

$$\mu_R \geq 107 \text{ packects/s}$$

With μ_R as obtained above, we get $A_{0R} = 0.935$, if we now consider R as a system with finite queue size. With such an offered load the queue size L (measured in number of packets) that guarantees a packet loss probability $\pi_p \leq \pi_0 = 10^{-6}$ is given by inverting formula (5.19):

$$A_0^L \leq \frac{\pi_0}{A_0 - A_0^2 + \pi_0 A_0^2}$$

resulting in:

$$L \geq 164$$

5.3.3.3   Load Balancing Between Two Output Links

A router connects a LAN to the Internet. The router has two interfaces, called I_1 and I_2, connected to two different links of capacity $C_1 = 100$ Mbit/s and $C_2 = 25$ Mbit/s, respectively. The packets from the LAN to the Internet arrive according to a Poisson process with average arrival rate $\lambda = 20{,}000$ packets/s and have sizes that are independent and identically distributed according to an exponential distribution with average $D = 500$ bytes $= 4000$ bits. Each interface has its own queue, which can be assumed of infinite size, and the router must balance the load between the two output interfaces. To achieve that, it forwards the packets randomly to interface I_1 with probability $P_1 = p$ and to interface I_2 with probability $P_2 = 1 - p$.

We need to find:

1. for which range of values of p the behavior of the queuing system is stable;
2. the value of p that minimizes the average time $\bar{\delta}$ spent in the router by the packets;
3. the minimum value $\delta_{\min}$.

Randomly splitting a Poisson arrival process results in multiple Poisson processes with proportionally reduced average arrival rates. Therefore, the router load balancing system can be studied as two $\mathcal{M}/\mathcal{M}/1$ queues in parallel, as illustrated in Fig. 5.12. The arrival and service rates for the two queues are:

$$\lambda_1 = \lambda p$$

$$\lambda_2 = \lambda(1 - p)$$

$$\mu_1 = \frac{C_1}{D} = 25000 \text{ packets/s}$$

$$\mu_2 = \frac{C_2}{D} = 6250 \text{ packets/s}$$

The two systems are stable if and only if:

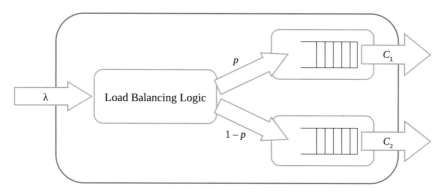

Fig. 5.12 Example of a router with two output interfaces and a load balancing logic to distribute traffic to both

$$\rho_1 = \frac{\lambda_1}{\mu_1} = \frac{20000}{25000} p < 1$$

and:

$$\rho_2 = \frac{\lambda_2}{\mu_2} = \frac{20000}{6250} (1 - p) < 1$$

leading to:

$$0.6875 < p < 1$$

The average time $\bar{\delta}$ spent in the router by a packet is the mean of the average times spent in the two interfaces:

$$\bar{\delta} = p\bar{\delta}_1 + (1 - p)\bar{\delta}_2 = \frac{p}{\mu_1 - \lambda p} + \frac{1 - p}{\mu_2 - \lambda(1 - p)} \tag{5.25}$$

The previous expression is plotted in Fig. 5.13 for $0.6875 < p < 1$. The curve shows a minimum in the range considered that can be found by setting the derivative of $\bar{\delta}$ to zero:

$$\frac{d\bar{\delta}}{dp} = \frac{\mu_1}{(\mu_1 - \lambda p)^2} - \frac{\mu_2}{(\mu_2 - \lambda(1 - p))^2} = 0 \tag{5.26}$$

Since in our case $\mu_1 = 4\mu_2$, replacing μ_1 with $4\mu_2$ in (5.26) and solving the equation for p, we obtain two possible solutions:

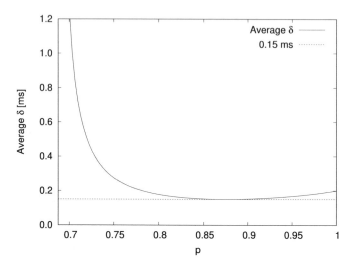

Fig. 5.13 $\bar{\delta}$ as a function p, with $0.6875 < p < 1$

$$p = \begin{cases} \dfrac{2}{3}\left(1 + \dfrac{\mu_2}{\lambda}\right) = 0.875 \\[2mm] 2\left(1 - \dfrac{3\mu_2}{\lambda}\right) = 0.125 \end{cases}$$

Only $p = 0.875$ is within the stability range of the system, which gives $\lambda_1 = 17{,}500$ packets/s and $\lambda_2 = 2500$ packets/s. Considering that interface I_1 is four times faster than I_2, it is reasonable to find that, to minimize the latency introduced by the router, almost all traffic must be forwarded to I_1. Yet, 12.5% of the packets must be forwarded to the slower interface to take advantage of the additional output channel and benefit from load balancing.

Finally, with $p = 0.875$ the minimum average time spent in the router is $\bar{\delta}_{min} = 0.15$ ms.

5.3.3.4 A Voice over IP Interconnection

Let us consider a link interconnecting the telephone networks of two different sites. The option is to implement a Voice over IP (VoIP) link using a data interconnection based on the Internet Protocol (IP). The behavior of the data flow resulting from the analog-to-digital conversion of a voice conversation is a well-known subject. Here let us recall some of the general characteristics. The traffic flow is usually bidirectional (*full duplex*) and almost symmetric. This means that each speaker speaks for almost half of the time of the phone call. When a speaker is talking, the voice is not continuous because of the natural pauses for breathing, word emphasizing, etc. Therefore, the digitization of the voice generates a number of data blocks, called *talkspurts*, which are interleaved with pauses.

Let us assume that talkspurts are carried by IP packets and that their lengths are independent and identically distributed with an exponential distribution of average $D = 22,528$ bits, plus an additional 10% overhead due to packet headers. Let us also assume that the talkspurts are generated according to a Poisson process with average arrival rate $\lambda_1 = 1$ packet/s. Indeed, all the packets carrying talkspurts generated by N active conversations will arrive at the router interconnecting the two sites, as an overall Poisson process with total arrival rate $\lambda = N\lambda_1$.

For the sake of comparison, let us consider an interconnection link between the two routers with a full duplex capacity equivalent to that of a standard E1 circuit-based digital telephony channel, i.e., $C = 2.048$ Mbit/s. This value takes into account that a single digitized phone call is allocated a full duplex capacity $C_0 = 64$ Kbit/s and, according to the European standard, an E1 channel can carry up to 30 phone calls (in principle $C/C_0 = 2048/64 = 32$, but two channels are reserved for signaling purposes).

The goal is to find:

- N_{max}: the maximum number of phone calls that can be simultaneously active between the two sites, so that the probability that a packet stays in the queue (considering only the packets that are actually queued) for more than $\varepsilon_0 = 200$ ms is less than $\pi_0 = 5\%$;
- π_q: the overall percentage of packets (considering any packet) that are queued for more than $\varepsilon_0 = 200$ ms, which we assume will not be useful to reconstruct the original voice with a digital-to-analog conversion because of an excessive jitter.

Since the problem is symmetrical, we limit our study to the output queue of one of the two routers, considering it as an $\mathcal{M}/\mathcal{M}/1$ system. The average service time is $\bar{\vartheta} = \frac{1}{\mu} = \frac{0.1D + D}{C} = 12.1$ ms.

For the $\mathcal{M}/\mathcal{M}/1$ system we know that:

$$\Pr\{\epsilon \leq \epsilon_0\} = F_\varepsilon(\varepsilon_0) = 1 - e^{-(\mu - \lambda)\varepsilon_0} \tag{5.27}$$

therefore:

$$1 - F_\varepsilon(\varepsilon_0) \leq \pi_0$$

$$e^{-(\mu - \lambda)\epsilon_0} \leq \pi_0$$

$$\lambda = N\lambda_1 \leq \mu + \frac{\ln(\pi_0)}{\epsilon_0} = 67.67 \text{ packets/s}$$

and finally:

$$N_{max} = 67$$

The resulting offered traffic and server utilization is $A_0 = \rho = 0.81$. Therefore, the probability that a packet containing a talkspurt is queued is:

$$\pi_r = \rho = 0.81$$

The exact probability that a talkspurt, once queued, is then delayed for more than 200 ms is:

$$\pi_{\varepsilon_0} = 1 - F_\varepsilon(\varepsilon_0) = e^{-(82.6-67)0.2} = 0.044$$

Therefore, the overall percentage of talkspurts that are delayed for more than 200 ms, considering any packet, is:

$$\pi_q = \pi_r \pi_{\varepsilon_0} = 0.036$$

It is interesting to note that the number of voice calls that can be supported by the packet-switched link is more than twice the number of calls supported by conventional circuit-based telephony. This is due to the low activity of the single voice source assumed in this example. As a matter of fact, the average bit rate of a single voice call is given by $\bar{C} = \lambda_1 \cdot D = 22.528$ Kbit/s, which is almost half the standard digital telephone circuit capacity. If the activity of the single source increases, obviously the total arrival rate λ also increases and N_{max} decreases.

5.3.3.5   Multiplexing Multi-Service Traffic

A University must connect the LANs of two remote sites with a dedicated data link. The traffic between the two sites can be classified into two types, but in both cases the packet arrival process can be considered Poisson and the packet size exponentially distributed with average $D = 1000$ bytes $= 8000$ bits. The traffic types are:

- scientific traffic, generated by research and teaching activities, with arrival rate $\lambda_1 = 1.6 \cdot 10^5$ packets/s;
- administrative traffic, generated by the activities of the administration offices, with arrival rate $\lambda_2 = 4 \cdot 10^4$ packets/s.

Two architectural alternatives must be considered:

1. to merge all traffic and deploy only one interconnection link between the two sites;
2. to keep the two traffic types separated, using two interconnection links in parallel, the former reserved to administrative traffic, the latter used for scientific traffic.

Let us assume that in both cases infinite queuing will be available at the router. The request is to dimension the link bit rate of the interfaces in the two cases mentioned above and identify the most effective alternative, subject to the following constraints:

- the link bit rate must be a multiple of the basic bit rate $C_0 = 10$ Mbit/s;
- for scientific traffic, the quality of service target is $\bar{\eta}_1 \leq 0.1$ ms;
- for administrative traffic, the quality of service target is $\bar{\varepsilon}_2 \leq 0.1$ ms.

Let us at first compute the average service rate, common to both types of traffic, considering a generic multiple N of the basic bit rate:

$$\mu = \frac{N C_0}{D} = \frac{N \cdot 10^7}{8 \cdot 10^3} = 1250 \, N \text{ packets/s}$$

Now let us consider the two possible alternatives, starting from the case without traffic separation. In this case, the overall packet arrival rate is $\lambda = \lambda_1 + \lambda_2 = 2 \cdot 10^5$ packets/s, which is offered to a single $\mathcal{M}/\mathcal{M}/1$ system, i.e.:

$$A_0 = \rho = \frac{\lambda}{\mu} = \frac{2 \cdot 10^5}{1250 \, N} = \frac{160}{N} \text{ E}$$

from which we know immediately that, to guarantee stability, $N > 160$, i.e., the link bit rate must be $C > 1.6$ Gbit/s.

Then let us find the value of N that satisfies the required quality of service for both traffic types. The value of N to choose must be the minimum one that satisfies both constraints on waiting times, i.e., $\bar{\eta} \leq 0.1$ ms and $\bar{\varepsilon} \leq 0.1$ ms. For scientific traffic we must choose N such that:

$$\bar{\eta} = \frac{1}{\mu} \frac{\rho}{1 - \rho} = \frac{1}{1250 \, N} \frac{160}{N - 160} \leq 10^{-4} \text{ s}$$

Rearranging the expression above gives the following quadratic inequality:

$$N^2 - 160 \, N - 1280 \geq 0$$

which is satisfied for $N \leq -7.64$ or $N \geq 167.64$. Of course only the latter is a valid solution, since N must be positive. Therefore, we can choose:

$$N = 168$$

Similarly, in case of administrative traffic we must choose N such that:

$$\bar{\varepsilon} = \frac{1}{\mu - \lambda} = \frac{1}{1250 \, N - 2 \cdot 10^5} \leq 10^{-4} \text{ s}$$

resulting in:

$$N \geq \frac{2.1 \cdot 10^5}{1250} = 168$$

This is exactly the same value chosen to satisfy the constrain on $\bar{\eta}$. Therefore, both quality of service targets are met with the following link bit rate:

$$C = 168\,C_0 = 1.68 \text{ Gbit/s}$$

Considering now the second architectural alternative, we have to study two separate $\mathcal{M}/\mathcal{M}/1$ systems and find the values of the multiples of C_0, i.e., N_1 and N_2, that satisfy the respective constraints. The traffic loads to the two queues are:

$$\rho_1 = \frac{\lambda_1}{\mu_1} = \frac{1.6 \cdot 10^5}{1250\,N_1} = \frac{128}{N_1} \quad \text{with } N_1 > 128$$

for scientific traffic and:

$$\rho_2 = \frac{\lambda_2}{\mu_2} = \frac{4 \cdot 10^4}{1250\,N_2} = \frac{32}{N_2} \quad \text{with } N_2 > 32$$

for administrative traffic, respectively.

The required link capacities for the two systems can then be obtained as follows. For scientific traffic we set:

$$\bar{\eta}_1 = \frac{1}{\mu_1}\frac{\rho_1}{1 - \rho_1} = \frac{1}{1250\,N_1}\frac{128}{N_1 - 128} \leq 10^{-4} \text{ s}$$

resulting in the following quadratic inequality:

$$N_1^2 - 128\,N_1 - 1024 \geq 0$$

whose only valid solution is $N_1 \geq 135.55$. We can then choose:

$$N_1 = 136$$

and dimension the corresponding link bit rate as:

$$C_1 = 1.36 \text{ Gbit/s}$$

For administrative traffic we set:

$$\bar{\varepsilon}_2 = \frac{1}{\mu_2 - \lambda_2} = \frac{1}{1250\,N_2 - 4 \cdot 10^4} \leq 10^{-4} \text{ s}$$

resulting in:

$$N_2 \geq \frac{5 \cdot 10^4}{1250} = 40$$

We can then choose:

$$N_2 = 40$$

and dimension the corresponding link bit rate as:

$$C_2 = 400 \text{ Mbit/s}$$

In summary, in the former case the University needs the equivalent of $N = 168$ channels at bit rate C_0, for a total of $C = 1.68$ Gbit/s, whereas in the latter case it needs the equivalent of $N = N_1 + N_2 = 176$ channels at bit rate C_0, for a total of $C = 1.76$ Gbit/s. Once again we see that aggregating traffic pays off in terms of system efficiency: aggregating all traffic provides a better capacity sharing effect, resulting in less overall capacity needed to satisfy the quality of service target of all traffic.

5.3.3.6 Queue Length Measured in Number of Bits

Let us consider the output interface of a router, with transmission bit rate $C = 100$ Mbit/s. Let us assume that packets arrive as a Poisson process with average arrival rate $\lambda = 8000$ packets/s, and that packets have random size exponentially distributed with average $D = 1250$ bytes $= 10{,}000$ bits. A queue is used to store packets arriving when the output link is busy transmitting another packet. The number of bits in the queue is the sum of the sizes measured in bits of all the queued packets.

We wish to calculate:

1. the average number of packets in the queue A_c and the corresponding average size of memory kept busy by the queued packets $\bar{b}$;
2. the probability density function of the amount of bits in the queue $f_{b/k}(x)$ when there are k packets in the system (i.e., $k - 1$ in the queue and 1 in service);
3. the unconditioned probability density function of the amount of bits in the queue $f_b(x)$;
4. the probability that more than $B = 32$ Kbits are in the queue, where 1 Kbit = 1024 bits).

We can find the average service time as:

$$\bar{\vartheta} = \frac{D}{C} = \frac{10^4}{10^8} = 0.1 \text{ ms}$$

and the traffic load as:

$$A_0 = \rho = \lambda \bar{\vartheta} = 0.8 \text{ E}$$

Then the average number of packets in the queue is:

$$A_c = \frac{\rho^2}{1 - \rho} = 3.2 \text{ E}$$

and the queue size in bits is simply given by the average number of packets times the average packet size in bits:

$$\bar{b} = A_c D = 31.25 \text{ Kbits}$$

Now, to find $f_{b/k}(x)$ we have to consider two possible cases:

- $k \leq 1$: in this case the amount of memory kept busy is 0, therefore $f_{b/k}(x) = \delta(x)$ (where $\delta(x)$ is the Dirac delta function) with probability $P_0 + P_1 = 1 - \rho^2 = 0.36$.
- $k \geq 2$: in this case the amount of bits in the queue is a random variable, equal to the sum of $k - 1$ (one of the k packets is in service) exponential random variables that are independent and identically distributed with average D. The probability density function can be calculated in a way similar to the one used to obtain Eq. (4.78), resulting in:

$$f_{b/k}(x) = \left(\frac{1}{D}\right)^{k-1} \frac{x^{k-2}}{(k-2)!} e^{-x/D}$$

The probability density function $f_b(x)$ can then be obtained as:

$$f_b(x) = \sum_{k=0}^{\infty} P_k f_{b/k}(x) = (P_0 + P_1)\delta(x) + \sum_{k=2}^{\infty} P_k f_{b/k}(x)$$

$$= (1 - \rho^2)\delta(x) + \sum_{k=2}^{\infty} (1 - \rho)\rho^k \left(\frac{1}{D}\right)^{k-1} \frac{x^{k-2}}{(k-2)!} e^{-x/D}$$

(5.28)

$$= (1 - \rho^2)\delta(x) + (1 - \rho)\frac{\rho^2}{D} e^{-x/D} \sum_{h=0}^{\infty} \left(\frac{\rho x}{D}\right)^h \frac{1}{h!}$$

$$= (1 - \rho^2)\delta(x) + \rho^2 \frac{(1 - \rho)}{D} e^{-(1-\rho)x/D}$$

which reveals that when the queue is not empty, i.e., with probability $\Pr(k \geq 2) = \rho^2$, the queue size measured in bits follows a negative exponential distribution with parameter $\frac{1-\rho}{D}$.

Finally, the probability distribution function of the queue size measured in bits can be obtained as:

$$F_b(x) = 1 - \rho^2 e^{-(1-\rho)x/D}$$

Now that we know $F_b(x)$, we can calculate the probability that the size of the queue exceeds a certain amount as:

$$\Pr\{b \geq B\} = 1 - F_b(B) = 0.8^2 e^{-\frac{0.2 \cdot 32768}{10000}} = 0.33$$

5.4   A Look at Some More General Cases

5.4.1   Poisson Arrivals May Be Fine, But What About Service Time? The $\mathcal{M}/\mathcal{G}/1$ Queue

As already briefly discussed in the introduction, the communication protocols used in packet switching networks usually set some lower and upper bounds to the packet length. There are several reasons for this. Although an extensive discussion on such an issue is beyond the scopes of this work, we can briefly mention that the main motivations are related to performance and implementation constraints. In general, in packet switching the signaling information is included in the header of the packet itself. In order to achieve a reasonable efficiency in the use of network resources, it is desirable that the amount of signaling information is kept less than the amount of user data carried in a packet. By setting the header length and a minimum overall packet length it is possible to keep the ratio between signaling and user data below a desired level. Moreover, protocol implementations require to store packets in memory for header processing. The related software and hardware functionality are by far easier and more reliable to implement if the packet size has defined boundaries.

It follows that, if packets have a minimum and maximum length, their service time will also have a minimum and maximum value, a characteristics which is in contrast with an exponential distribution that, by definition, requires that the service time may take values between 0 and infinity. Moreover, besides the minimum and maximum size limits, the packet length may have some specific characteristics. For instance, some protocols require that all packets have the same fixed size, or for some reason a few specific sizes in the set of possible values are more frequent.

Based on the previous discussion, in general we can say that the assumption of exponential packet sizes is used as a first-level approximation, but it would be by far more realistic to consider a model that allows to take into account a more general packet size distribution. This is the scope of this section, where we consider a single server queuing system with Poisson arrivals and general service times, not necessarily memoryless. This is called the $\mathcal{M}/\mathcal{G}/1$ queuing system, which cannot be studied as a Markov chain as in the $\mathcal{M}/\mathcal{M}/1$ system case. As a matter of fact, a full analysis of the $\mathcal{M}/\mathcal{G}/1$ queue is rather complex and, for this reason, we will not delve into all the details. Nonetheless, an analysis of the average performance metrics is possible and still quite useful.

5.4.1.1 The Average Waiting Time $\bar{\eta}$ with FIFO Scheduling

At first, we assume a FIFO scheduling policy for an $\mathcal{M}/\mathcal{G}/1$ system. We also assume that we know:

- the average service time $\bar{\vartheta}$;
- the variance of the service time $\sigma_{\vartheta}^2 = E[\vartheta^2] - \bar{\vartheta}^2$.

Now let us consider what happens in general to a generic customer entering the system. The customer must wait to be served for a time η that is given by the sum of two different components:

$$\eta = T' + T'' \tag{5.29}$$

where:

- T' is the residual service time ζ of the customer being served, if any;
- T'' is the sum of the service times of the k_c customers already waiting in the queue before the customer considered, if any.

Obviously, we can consider the average values in Eq. (5.29) and get:

$$\bar{\eta} = E[T'] + E[T''] \tag{5.30}$$

We can calculate the two average components rather easily. Let us recall that for a single server system it is always $1 - P_0 = \rho$. When a customer arrives, it finds the server busy with probability $1 - P_0$ and free with probability P_0. Therefore:

$$T' = \begin{cases} 0 & \text{with probability } P_0 \\ \zeta & \text{with probability } 1 - P_0 \end{cases}$$

and then:

$$E[T'] = (1 - P_0)\bar{\zeta} + 0 \cdot P_0 = \rho\bar{\zeta} \tag{5.31}$$

where $\bar{\zeta}$ is the average residual general service time obtained in Eq. (2.53)

As for T'', each of the k_c customers already waiting in the queue will keep the server busy for a specific service time ϑ. Such service time is independent of the number of customers in the queue and of the specific customer considered. Therefore, k_c and ϑ can be considered as two independent random variables, and the average value of their product is the product of their average values. Since $E[k_c] = A_c$, we can then write:

$$E[T''] = E[k_c\vartheta] = A_c\bar{\vartheta} = \lambda\bar{\eta}\bar{\vartheta} = \rho\bar{\eta} \tag{5.32}$$

Therefore, Eq. (5.30) turns out as follows:

$$\bar{\eta} = \rho\bar{\zeta} + \rho\bar{\eta} \tag{5.33}$$

which can be solved for $\bar{\eta}$ to get:

$$\bar{\eta} = \bar{\zeta}\frac{\rho}{1-\rho} = \frac{\rho E[\vartheta^2]}{2(1-\rho)\bar{\vartheta}} = \frac{\lambda E[\vartheta^2]}{2(1-\rho)} = \frac{\lambda(\sigma_\vartheta^2 + \bar{\vartheta}^2)}{2(1-\rho)} \tag{5.34}$$

It is interesting to note that $\bar{\eta} = \bar{\zeta}\frac{\rho}{1-\rho}$ is very similar to Eq. (5.11), with the only difference that $\bar{\zeta}$ replaces $\bar{\vartheta}$. As a matter of fact, in case of exponential service time the memoryless property leads to $\bar{\zeta} = \bar{\vartheta}$, and therefore the two equations give the same result.

Starting from Eq. (5.34) we can obtain all the other relevant average quantities for the $\mathcal{M}/\mathcal{G}/1$ queue, i.e.:

$$\bar{\delta} = \bar{\vartheta} + \bar{\eta} = \bar{\vartheta} + \frac{\lambda E[\vartheta^2]}{2(1-\rho)} = \bar{\vartheta} + \frac{\rho^2 + \lambda^2\sigma_\vartheta^2}{2\lambda(1-\rho)} = \frac{2\rho - \rho^2 + \lambda^2\sigma_\vartheta^2}{2\lambda(1-\rho)} \tag{5.35}$$

from which we have:

$$A = \frac{2\rho - \rho^2 + \lambda^2\sigma_\vartheta^2}{2(1-\rho)} = \rho + \frac{\rho^2 + \lambda^2\sigma_\vartheta^2}{2(1-\rho)} \tag{5.36}$$

and finally:

$$A_c = \frac{\lambda^2 E[\vartheta^2]}{2(1-\rho)} = \frac{\rho^2 + \lambda^2\sigma_\vartheta^2}{2(1-\rho)} \tag{5.37}$$

Equation (5.36) is called the *Pollaczek–Khintchine average value formula*, from the names of the two scientists that first derived it. The behavior of A as a function of σ_ϑ^2 is shown in Fig. 5.14, which clearly shows that, for given values of λ and μ, the larger σ_ϑ^2, the larger A, and consequently the larger the waiting time. In other words, a service time with larger variance results in a larger number of queued customers and in a larger waiting time.

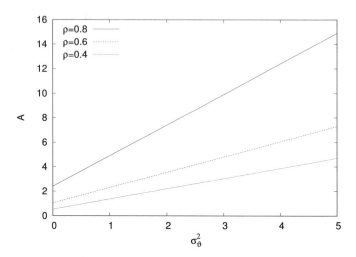

Fig. 5.14 $\mathcal{M}/\mathcal{G}/1$: A as a function σ_ϑ^2, for $\lambda = 1$ and varying ρ as a parameter

5.4.1.2 Some Relevant Cases of the $\mathcal{M}/\mathcal{G}/1$ Queue

In this paragraph we briefly review some results for relevant cases that fit into the $\mathcal{M}/\mathcal{G}/1$ model. In particular, we find that the results already known obtained for the $\mathcal{M}/\mathcal{M}/1$ queue can be also found as a specific case of the $\mathcal{M}/\mathcal{G}/1$ system. Then we study the $\mathcal{M}/\mathcal{D}/1$ system with deterministic service time, the $\mathcal{M}/\mathcal{U}/1$ with uniform service time, the $\mathcal{M}/\mathcal{E}_r/1$ with Erlang-r service time, and the $\mathcal{M}/\mathcal{P}\text{areto}/1$ with Pareto service time.

In case of service time with exponential distribution, since $\bar{\vartheta} = 1/\mu$ and $\sigma_\vartheta^2 = 1/\mu^2$, we can calculate:

$$A = \frac{2\rho - \rho^2 + \lambda^2 \sigma_\vartheta^2}{2(1-\rho)} = \frac{2\rho - \rho^2 + \rho^2}{2(1-\rho)} = \frac{\rho}{1-\rho}$$

$$A_c = \frac{\rho^2 + \lambda^2 \sigma_\vartheta^2}{2(1-\rho)} = \frac{\rho^2}{1-\rho}$$

and:

$$\bar{\eta} = \frac{\rho^2}{\lambda(1-\rho)} = \bar{\vartheta}\frac{\rho}{1-\rho}$$

which are exactly the results we found in Sect. 5.3.1.

In case of the $\mathcal{M}/\mathcal{D}/1$ system, the service time is fixed and therefore $\vartheta = \vartheta_0$ and $\sigma_\vartheta^2 = 0$. It follows that:

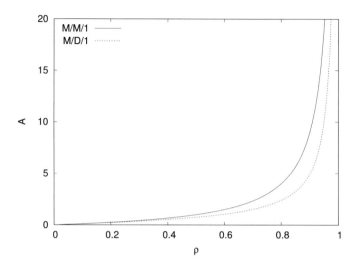

Fig. 5.15 A as a function of ρ comparing the $\mathcal{M}/\mathcal{D}/1$ and $\mathcal{M}/\mathcal{M}/1$ systems

$$A = \frac{2\rho - \rho^2 + \lambda^2 \sigma_\vartheta^2}{2(1-\rho)} = \frac{2\rho - \rho^2}{2(1-\rho)} = \frac{\rho}{1-\rho} - \frac{\rho^2}{2(1-\rho)}$$

$$A_c = \frac{\rho^2 + \lambda^2 \sigma_\vartheta^2}{2(1-\rho)} = \frac{\rho^2}{2(1-\rho)}$$

and:

$$\bar{\eta} = \frac{\rho^2}{2\lambda(1-\rho)} = \vartheta_0 \frac{\rho}{2(1-\rho)}$$

If we compare these formulas with the corresponding ones for the $\mathcal{M}/\mathcal{M}/1$ system, it is clear that the values of A, A_c and $\bar{\eta}$ are smaller in the deterministic case. In particular, we see that $\bar{\eta}_{\mathcal{M}/\mathcal{D}/1} = \frac{1}{2}\bar{\eta}_{\mathcal{M}/\mathcal{M}/1}$. This is also shown in Fig. 5.15, and is in line with the general comment that queuing performance gets worse with larger variance of the service time. Obviously, in case of deterministic service time the queuing performance will be the best because the variance of the service time is zero.

With the method of the z-transform described in the appendix, it is possible to calculate the steady state probabilities of the $\mathcal{M}/\mathcal{D}/1$ system:

$$P_k = \begin{cases} 1 - \rho & k = 0 \\ (1-\rho)(e^\rho - 1) & k = 1 \\ (1-\rho)\sum_{i=1}^{k}(-1)^{k-i}e^{i\rho}\left[\frac{(i\rho)^{k-i}}{(k-i)!} + \frac{(i\rho)^{k-i-1}}{(k-i-1)!}\right] & k > 1 \end{cases} \tag{5.38}$$

It is worth noting that in this case the probability distribution of the time spent in the system and of the time spent in the queue can be calculated directly from the steady state probabilities. In fact, a customer that finds k customers in the system when joining the queue, including the one currently being served, is going to stay in the system on average for $\bar{\delta}_k = (k + \frac{1}{2})\vartheta_0$, and is going to wait in the queue on average for $\bar{\eta}_k = (k - \frac{1}{2})\vartheta_0$, since the service time for each customer in the queue is fully deterministic and the average residual service time is $\frac{\vartheta_0}{2}$.

For a $\mathcal{M}/\mathcal{U}/1$ system, assuming the service time uniformly distributed between 0 and a maximum value ϑ_M, it follows that $\bar{\vartheta} = \frac{\vartheta_M}{2}$ and $\sigma_\vartheta^2 = \frac{\vartheta_M^2}{12}$. Therefore:

$$A = \frac{2\rho - \rho^2 + \lambda^2\sigma_\vartheta^2}{2(1-\rho)} = \frac{\rho}{1-\rho} - \frac{\frac{2}{3}\rho^2}{2(1-\rho)}$$

$$A_c = \frac{\rho^2 + \lambda^2\sigma_\vartheta^2}{2(1-\rho)} = \frac{\frac{4}{3}\rho^2}{2(1-\rho)}$$

and:

$$\bar{\eta} = \frac{\frac{4}{3}\rho^2}{2\lambda(1-\rho)} = \frac{\vartheta_M}{2}\frac{\frac{4}{3}\rho}{2(1-\rho)} = \vartheta_M\frac{\rho}{3(1-\rho)}$$

The $\mathcal{M}/\mathcal{E}_r/1$ system is a rather interesting case that provides some additional insights on the performance analysis of the $\mathcal{M}/\mathcal{G}/1$ queue. Let us imagine that the service requested by the customers is composed of a series of phases offered by r subsequent stations, according to the following assumptions:

- the service time in each the r phases is exponentially distributed with average duration $\bar{\vartheta}_0$;
- the service times in the r phases are independent of each other;
- a new customer may be served if and only if no service station is busy (i.e., when the full service of any previous customer is completed).

Such behavior of the service time can be modeled as a random variable with an Erlang distribution of grade r, as defined in Sect. 2.3.4. Under these assumptions, the average service time per customer is $\bar{\vartheta} = r\,\bar{\vartheta}_0$ and the variance is $\sigma_\vartheta^2 = r\,\bar{\vartheta}_0^2$. Therefore:

$$A = \frac{2\rho - \rho^2 + \lambda^2\sigma_\vartheta^2}{2(1-\rho)} = \frac{\rho}{1-\rho} - \frac{\left(1 - \frac{1}{r}\right)\rho^2}{2(1-\rho)}$$

$$A_c = \frac{\rho^2 + \lambda^2\sigma_\vartheta^2}{2(1-\rho)} = \frac{\left(1 + \frac{1}{r}\right)\rho^2}{2(1-\rho)}$$

and:

$$\bar{\eta} = \frac{\left(1 + \frac{1}{r}\right)\rho^2}{2\lambda(1-\rho)} = \bar{\vartheta}\frac{\left(1 + \frac{1}{r}\right)\rho}{2(1-\rho)} = \bar{\vartheta}_0\frac{(r+1)\rho}{2(1-\rho)}$$

Recalling from Sect. 2.3.4 that an Erlang-r distribution becomes a simple exponential for $r = 1$ and converges to a deterministic one for $r \to \infty$, we expect that an $\mathcal{M}/\mathcal{E}_r/1$ system behaves as an $\mathcal{M}/\mathcal{M}/1$ for $r = 1$ and tends to the $\mathcal{M}/\mathcal{D}/1$ for very large values of r. This is confirmed if we compute A in the two extreme cases, i.e.:

$$A(r = 1) = \frac{\rho}{1-\rho}$$

and:

$$\lim_{r \to \infty} A(r) = \frac{\rho}{1-\rho} - \frac{\rho^2}{2(1-\rho)}$$

Assuming a service time following a Pareto distribution as defined in Sect. 2.3.5, the $\mathcal{M}/\mathcal{P}\text{areto}/1$ system can be studied using formula (5.36) only under the condition that the variance is finite, i.e., when the shape parameter is $\alpha > 2$. In this case we have:

$$\bar{\vartheta} = \frac{\alpha \, t_0}{\alpha - 1} \quad \text{and} \quad \sigma_{\bar{\vartheta}}^2 = \left(\frac{t_0}{\alpha - 1}\right)^2 \frac{\alpha}{\alpha - 2} = \frac{\bar{\vartheta}^2}{\alpha(\alpha - 2)}$$

where $t_0 > 0$ is the scale parameter. Therefore:

$$A = \frac{2\rho - \rho^2 + \lambda^2\sigma_{\bar{\vartheta}}^2}{2(1-\rho)} = \frac{\rho}{1-\rho} - \frac{\left(1 - \frac{1}{\alpha(\alpha-2)}\right)\rho^2}{2(1-\rho)}$$

$$A_c = \frac{\rho^2 + \lambda^2\sigma_{\bar{\vartheta}}^2}{2(1-\rho)} = \frac{\left(1 + \frac{1}{\alpha(\alpha-2)}\right)\rho^2}{2(1-\rho)}$$

and:

$$\bar{\eta} = \frac{\left(1 + \frac{1}{\alpha(\alpha-2)}\right)\rho^2}{2\lambda(1-\rho)} = \bar{\vartheta}\frac{\left(1 + \frac{1}{\alpha(\alpha-2)}\right)\rho}{2(1-\rho)} = \frac{\alpha \, t_0}{\alpha - 1}\frac{\left(1 + \frac{1}{\alpha(\alpha-2)}\right)\rho}{2(1-\rho)}$$

Figure 5.16 compares how A increases with ρ for some of the $\mathcal{M}/\mathcal{G}/1$ systems considered above. As expected, for a given load ρ the average traffic in the system decreases when the variance of the service time decreases, with the Pareto service

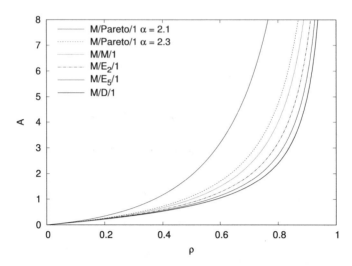

Fig. 5.16 A as a function of ρ comparing Pareto, exponential, Erlang-r and deterministic service time distributions

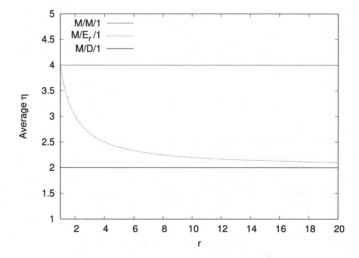

Fig. 5.17 $\mathcal{M}/\mathcal{E}_r/1$ system: average waiting time with $\bar{\vartheta} = 1$ e $\rho = 0.8$ as a function of r, compared to that of the $\mathcal{M}/\mathcal{M}/1$ and $\mathcal{M}/\mathcal{D}/1$ systems

time being the worst case when α is close to 2 and the Erlang-r cases appearing as intermediate between the exponential and deterministic service time distributions.

Finally, the behavior of the average queuing time $\bar{\eta}$ in a $\mathcal{M}/\mathcal{E}_r/1$ system as a function of r is shown in Fig. 5.17, with $\bar{\vartheta} = 1$ e $\rho = 0.8$. Once again it is apparent how the system performance lies between that of an $\mathcal{M}/\mathcal{M}/1$ and an $\mathcal{M}/\mathcal{D}/1$.

5.4.2   Packet Switching and Quality of Service: When One Pipe Does Not Fit All

Up to now we have considered mainly queuing systems with a FIFO scheduling policy. This is indeed the most common policy used in many practical applications, and is also what we, as human beings, tend to consider the most reasonable and fairest policy. Nonetheless, in many situations of practical interest non FIFO scheduling is appropriate and necessary. This is particularly true in packet switching, where different traffic flows related to different services or contents may have different QoS requirements, and should therefore be treated differently when queued. In this section we briefly review a couple of simple models, closely related to the $\mathcal{M}/\mathcal{G}/1$ queue but assuming other than FIFO scheduling policies, in particular:

- priority queuing;
- shortest job next (SJN).

5.4.2.1   Priority Queuing

The scenario considered here is that of a queuing system behaving as follows:

- arrivals can be split into R traffic flows (or classes), each with Poisson arrivals with average rate $\lambda_r, r = 1, \ldots, R$;
- the service time of traffic flow r is ϑ_r, with average value $\bar{\vartheta}_r = 1/\mu_r$;
- the scheduling in the queue works as follows:

  - packets from the same class are scheduled to be served according to a FIFO policy;
  - traffic flows with lower index have higher priority, therefore packets from class r joining the queue are served before packets from any class $j > r$ and after packets from any class $i < r$;
  - while a packet is waiting in the queue, if there is an arrival of a packet with higher priority, such packet overcomes the one that is waiting and is scheduled to be served before it.

What described above is the basic concept of priority scheduling. To fully clarify the queue behavior, there is an additional issue to be considered, i.e., the behavior of the server when a new packet arrives. We consider two possible cases:

non-preemptive priority: the service is fully decoupled from the priority scheduling, meaning that when a new packet arrives and finds the server busy, the server will continue to serve the current packet as planned, and only when the current service is completed the priority scheduling will be applied to serve a new packet from the queue;

preemptive priority: the service must consider the new arriving packets and, in case a higher priority packet arrives, it stops serving the lower priority packet and starts serving the higher priority one.

Obviously, the latter case gives an additional advantage to the priority packets that not only advance in the queue but also experience more limited waiting times. Nonetheless, preemption implies that the interrupted packet is partially served, and the additional issue of what to do next arises. Should we park the packet that was preempted aside and resume its service when possible, or should we put it back in the queue to start again the service from scratch later on? If we focus on networking scenarios, we find that resuming the service does not make much sense. In a packet-switched network a packet is a sort of atomic entity that must be transmitted altogether to be meaningful, unless specific fragmentation techniques are allowed by the involved protocols that must be in any case applied before the packet transmission begins. If we transmit a packet only partially, it would carry incomplete information that could not satisfy protocol-specific integrity controls and then would likely be discarded at the next network node. Therefore, for the scenarios relevant to this section, the preemptive priority is not considered efficient. We then focus on non-preemptive priority scheduling only. The goal is to find the average waiting time for the packets of each priority class.

This problem can be solved in a way similar to what was done in Sect. 5.4.1.1. When a packet of class r joins the queue, its waiting time consists of three contributions:

- T' is the residual service time of the packet currently being served, if any;
- T_i'' is the service time of all the packets already waiting in the queue belonging to class $i \leq r$, if any;
- T_i''' is the service time of all the packets belonging to class $i < r$ that join the queue during the waiting time of the packet we are considering, if any.

Considering a non-preemptive priority scheduling mechanism, we can write:

$$\eta_1 = T' + T_1''$$

$$\eta_r = T' + \sum_{i=1}^{r} T_i'' + \sum_{i=1}^{r-1} T_i''' \qquad r = 2, \ldots, R \tag{5.39}$$

In terms of average values:

$$\bar{\eta}_1 = E[T'] + E[T_1'']$$

$$\bar{\eta}_r = E[T'] + \sum_{i=1}^{r} E[T_i''] + \sum_{i=1}^{r-1} E[T_i'''] \qquad r = 2, \ldots, R \tag{5.40}$$

The terms of this formula can be calculated as follows:

- The average residual service time of a packet belonging to priority class i is given by Eq. (2.53), i.e.:

$$\bar{\zeta}_i = E[\vartheta_i^2]/2\bar{\vartheta}_i$$

Moreover, the probability that in a random time instant a packet of class i is currently being served is given by the server utilization by class i packets, i.e., $\rho_i = \lambda_i/\mu_i$. Therefore:

$$E[T'] = \sum_{i=1}^{R} \rho_i \frac{E[\vartheta_i^2]}{2\bar{\vartheta}_i} = \frac{1}{2}\sum_{i=1}^{R} \lambda_i E[\vartheta_i^2]$$

- $E[T_i'']$ can be found by applying Little's Theorem to traffic flow i. The average number of packets of class i in the queue is $A_{ci} = \lambda_i \bar{\eta}_i$. Since each of these packets requires an average service time $\bar{\vartheta}_i$, it follows that:

$$E[T_i''] = \lambda_i \bar{\eta}_i \bar{\vartheta}_i = \rho_i \bar{\eta}_i$$

- $E[T_i''']$ can be calculated considering the number of packets of class i arriving in an average time period equal to $\bar{\eta}_r$, which is $\lambda_i \bar{\eta}_r$. Since each of these packets requires an average service time $\bar{\vartheta}_i$, it follows that:

$$E[T_i'''] = \lambda_i \bar{\eta}_r \bar{\vartheta}_i = \rho_i \bar{\eta}_r$$

Putting everything together, for the generic class $r = 1, \ldots, R$ we obtain:[2]

$$\bar{\eta}_r = \frac{1}{2}\sum_{i=1}^{R} \lambda_i E[\vartheta_i^2] + \sum_{i=1}^{r} \rho_i \bar{\eta}_i + \sum_{i=1}^{r-1} \rho_i \bar{\eta}_r = \frac{1}{2}\sum_{i=1}^{R} \lambda_i E[\vartheta_i^2] + \sum_{i=1}^{r-1} \rho_i \bar{\eta}_i + \sum_{i=1}^{r} \rho_i \bar{\eta}_r$$

which can be solved for $\bar{\eta}_r$ leading to the following result:

$$\bar{\eta}_r = \frac{\frac{1}{2}\sum_{i=1}^{R} \lambda_i E[\vartheta_i^2] + \sum_{i=1}^{r-1} \rho_i \bar{\eta}_i}{1 - \sum_{i=1}^{r} \rho_i} \tag{5.41}$$

Starting from Eq. (5.41) we can also calculate the average time spent in the queue by a packet of class r as $\bar{\delta}_r = \bar{\vartheta}_r + \bar{\eta}_r$.

Formula (5.41) can be rewritten in a more compact form by defining:

---

[2] The results for class $r = 1$ can be obtained assuming $\sum_{i=1}^{0} \rho_i \bar{\eta}_i = 0$.

$$\bar{\vartheta}_P = \frac{1}{2} \sum_{i=1}^{R} \lambda_i E[\vartheta_i^2]$$

and

$$S_0 = 0 \qquad S_r = \sum_{i=1}^{r} \rho_i \qquad r = 1, \ldots, R$$

Doing so we can write:

$$\bar{\eta}_1 = \frac{\bar{\vartheta}_P}{1 - S_1}$$

$$\bar{\eta}_2 = \frac{\bar{\vartheta}_P + \rho_1 \bar{\eta}_1}{1 - S_2} = \frac{\bar{\vartheta}_P}{(1 - S_1)(1 - S_2)}$$

and, in general:

$$\bar{\eta}_r = \frac{\bar{\vartheta}_P}{(1 - S_{r-1})(1 - S_r)} \qquad r = 1, \ldots, R \tag{5.42}$$

which is called *Cobham's formula*.[3] This expression of the average waiting time of packet of class r shows the effect of the traffic share of the different classes. Any class $i > r$ has an impact only on the weighted average residual service time $\bar{\vartheta}_P$, due to the non-preemptive scheduling considered. Then the average waiting time is increased by two factors that depend only on the load share of classes up to r. Since

---

[3]Cobham's formula can be demonstrated by induction. It is straightforward to see that it is valid for $r = 1$. Then, assuming it is valid for a generic $r = j$, we prove that it also holds for $r = j + 1$. From (5.41) for $r = j$ we get:

$$\sum_{i=1}^{j-1} \rho_i \bar{\eta}_i = (1 - S_j)\bar{\eta}_j - \bar{\vartheta}_P$$

This can be replaced in formula (5.41) considering $r = j + 1$, finding that:

$$\bar{\eta}_{j+1} = \frac{\bar{\vartheta}_P + \sum_{i=1}^{j} \rho_i \bar{\eta}_i}{1 - S_{j+1}} = \frac{\bar{\vartheta}_P + \sum_{i=1}^{j-1} \rho_i \bar{\eta}_i + \rho_j \bar{\eta}_j}{1 - S_{j+1}} = \frac{\bar{\vartheta}_P + (1 - S_j)\bar{\eta}_j - \bar{\vartheta}_P + \rho_j \bar{\eta}_j}{1 - S_{j+1}}$$

and therefore:

$$\bar{\eta}_{j+1} = \frac{1 - S_j + \rho_j}{1 - S_{j+1}} \bar{\eta}_j = \frac{1 - S_{j-1}}{1 - S_{j+1}} \frac{\bar{\vartheta}_P}{(1 - S_{j-1})(1 - S_j)} = \frac{\bar{\vartheta}_P}{(1 - S_j)(1 - S_{j+1})}$$

$\bar{\vartheta}_P$ does not depend on r, considering how the terms S_r are defined, it is easy to see that $\bar{\eta}_r$ is larger when r increases.

5.4.2.2   Shortest Job Next (SJN) Scheduling

Let us consider an $\mathcal{M}/\mathcal{G}/1$ system with a special scheduling policy that gives priority to the packets depending on their length: short packets overcome longer ones in the queue without preemption. This case is very similar to the non-preemptive priority scheduling presented above, with the difference that here the priority classes are as many as the possible values of the packet length, in theory also infinite. Therefore, we can study this problem using the same approach as in Sect. 5.4.2.1 taking into account that priority class r includes all packets with service time $(r-1)\Delta\vartheta \leq \vartheta \leq i\Delta\vartheta$, where $\Delta\vartheta$ is a suitable fixed time interval used as a reference to define the different classes based on the service time.

If λ is as usual the total average packet arrival rate, then the arrival rate of class r is:

$$\lambda_r = \lambda \Pr\{(r-1)\Delta\vartheta \leq \vartheta \leq r\Delta\vartheta\}$$

$$= \lambda\left[F_\vartheta(r\Delta\vartheta) - F_\vartheta((r-1)\Delta\vartheta)\right] = \lambda \int_{(r-1)\Delta\vartheta}^{r\Delta\vartheta} f_\vartheta(t)dt \qquad (5.43)$$

Moreover, the average load of class r is:

$$\rho_r = \lambda_r \bar{\vartheta}_r$$

where $\bar{\vartheta}_r$ is the average service time of packets in class r, i.e.:

$$\bar{\vartheta}_r = \frac{\int_{(r-1)\Delta\vartheta}^{r\Delta\vartheta} t f_\vartheta(t)dt}{\int_{(r-1)\Delta\vartheta}^{r\Delta\vartheta} f_\vartheta(t)dt}$$

Therefore, we can re-write Cobham's formula in Eq. (5.42) as:

$$\bar{\eta}_r = \frac{\bar{\vartheta}_P}{(1 - S_{r-1})(1 - S_r)}$$

where the weighted average residual service time $\bar{\vartheta}_P$ must be computed over all traffic classes, which in this case means for any possible value of the service time, resulting in:

$$\bar{\vartheta}_P = E[T'] = \rho\bar{\zeta} = \rho\frac{E[\vartheta^2]}{2\bar{\vartheta}} = \frac{\lambda E[\vartheta^2]}{2}$$

If we imagine to have infinite priority classes, each associated with a possible value of the packet service time that is a continuous random variable, then we can apply the class definition given above considering infinitesimal intervals, i.e., assuming that $\Delta\vartheta \to 0$. The generic priority class is now identified by a specific value of ϑ and not by r any more, and the quantities used so far become as follows:

$$\lambda_r \to \lambda(\vartheta) = \lambda f_\vartheta(\vartheta)\, d\vartheta$$

$$\rho_r \to \rho(\vartheta) = \lambda(\vartheta)\vartheta = \lambda\vartheta f_\vartheta(\vartheta)\, d\vartheta$$

$$S_r, S_{r-1} \to S(\vartheta) = \int_0^\vartheta \lambda t f_\vartheta(t)\, dt$$

$$\bar{\eta}_r \to \bar{\eta}(\vartheta) = \frac{\bar{\vartheta}_P}{(1 - S(\vartheta))^2} = \frac{\lambda E[\vartheta^2]}{2(1 - S(\vartheta))^2}$$

The average waiting time for a generic user can be calculated as an average of $\bar{\eta}(\vartheta)$ over all possible values of ϑ.

$$\bar{\eta} = \int_0^\infty \bar{\eta}(\vartheta) f_\vartheta(\vartheta)\, d\vartheta$$

5.4.2.3 Kleinrock's Conservation Law

Let us consider a non-preemptive priority scheduling system with total workload $\rho = \sum_{i=1}^R \rho_i$. When compared to a FIFO scheduling with load ρ, the priority scheduling does not alter the total amount of work that the server has to perform, given that the overall utilization remains the same. The priority scheduling changes only the order in which the packets are served, but at some point all of the will be served. Based on this observation an interesting result can be obtained, known as *Kleinrock's conservation law*.

Let us define the *unfinished workload* $U(t)$ as the amount of time required to completely serve all the packets that are in the system at time t. This is equivalent to measuring the time interval, starting at t, in which the server will be busy assuming that no further arrival occurs after t. $U(t)$ can then be calculated as the sum of:

1. the residual service time of the packet being served at time t, if any;
2. the sum of the service times of all the packets waiting in the queue at time t, if any.

$U(t)$ is a quantity that can be either zero or positive, and which decreases at a constant rate, i.e., the service speed. Moreover, it increases instantaneously at each new arrival by a quantity equal to the service time of the newly arrived packet, as shown graphically in the example of Fig. 5.18.

The most important observation is that $U(t)$ is independent of the queuing discipline, as long as the total workload is preserved. A queuing system that

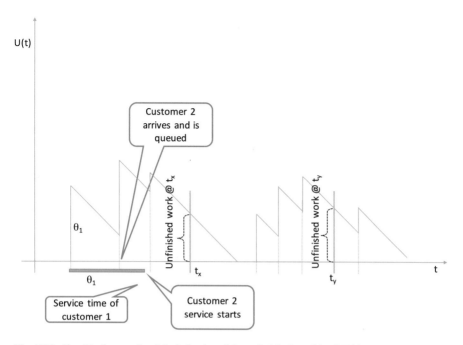

Fig. 5.18 Graphical example of the behavior of the unfinished workload $U(t)$

guarantees this principle is called a *work conserving* system. For this to be true we must assume that:

- no packet leaves the system unless it has been fully served, which is part of the assumption of a non-preemptive scheduling;
- new arrivals introduce in the system an additional workload equal to the service time of the new packet, without modifying the service time of other packets already in the system.

Both assumptions are fairly intuitive and reasonable in the case of packet switching systems, as those considered here.

Under the assumption of a work conserving system subject to arrivals from R traffic classes, it is possible to prove that:

$$\sum_{i=1}^{R} \rho_i \bar{\eta}_i = \bar{\vartheta}_P \frac{\rho}{1 - \rho} \tag{5.44}$$

Indeed, if we consider a system with R priority classes, using the same notation as in Eq. (5.40), we can write:

$$U(t) = T' + \sum_{i=1}^{R} T_i''$$

and the average is:

$$\bar{U} = E[U(t)] = E[T'] + \sum_{i=1}^{R} E[T_i''] = \bar{\vartheta}_P + \sum_{i=1}^{R} \rho_i \bar{\eta}_i \qquad (5.45)$$

Since $U(t)$ does not depend on the queuing discipline, we can calculate the previous quantity for an $\mathcal{M}/\mathcal{G}/1$ system with FIFO scheduling. Thanks to the PASTA property, the average value of $U(t)$ coincides with the average waiting time in the queue $\bar{\eta}$ of a generic packet arriving at the system. Therefore:[4]

$$\bar{U} = \bar{\eta} = \bar{\zeta} \frac{\rho}{1-\rho} = \frac{\bar{\vartheta}_P}{1-\rho} \qquad (5.46)$$

Combining Eqs. (5.45) and (5.46) we obtain the expression of Kleinrock's conservation law (5.44), which tells us that the total weighted average waiting time in a work conserving system is constant. Therefore, if we give an advantage to some packets making them wait less, this happens at the expenses of some other packets that have to wait more. This is a rather intuitive result, but Kleinrock's conservation law allows us to quantitatively assess this phenomenon, showing that the effect of any change in the service scheduling order depends on the relative load shares of the different traffic classes.

5.4.3 Examples and Case Studies

5.4.3.1 Stop-and-Wait with Errors

A data link protocol is implemented with a stop-and-wait flow control algorithm. The transmitter is allowed to send one data frame at a time and has to wait for the acknowledgement from the receiver of the correct frame reception before transmitting the next data frame. Under normal operating conditions, the acknowledgement message (ACK) is received after $T = 2.5$ ms since the beginning of the transmission of the data frame. If no ACK is received within $T_o = 3$ ms since the beginning of the transmission of the data frame, it is assumed that something went wrong in the channel and the data frame is immediately re-transmitted.

---

[4]In an $\mathcal{M}/\mathcal{G}/1$ system subject to a single traffic class, the weighted average residual service time is given by $\bar{\vartheta}_P = \bar{\zeta}\rho$.

The frame error probability is $P_F = 0.1$, which is the same for all frames and is independent from frame to frame. We want to find:

1. the data frame service time ϑ_k, i.e., the time needed to send a single data frame successfully, measured from the beginning of the transmission of the data frame to the time the ACK is received, when k frame transmission errors happen;
2. the probability of the data frame service time ϑ_k;
3. the average data frame service time $\bar{\vartheta}$.

Considering the frame arrivals distributed as a Poisson process with average arrival rate λ frames/s and assuming an infinite queue size at the transmitter, which acts as server, we also want to obtain:[5]

4. the maximum arrival rate λ_M that guarantees a stable behavior of the queue;
5. the average data frame waiting time for $\lambda = 0.8\lambda_M$.

The frame transmission time ϑ_k when k errors happen is given by the sum of k transmissions with error followed by one last successful transmission, i.e.:

$$\vartheta_k = kT_o + T$$

Given the frame error probability P_F, the probability of having k independent errors followed by one success is:

$$p_k = P_F^k(1 - P_F)$$

which is also the probability of service time ϑ_k. Then we can obtain the average data frame service time as:

$$\bar{\vartheta} = \sum_{k=0}^{\infty} \vartheta_k p_k = T + T_o(1 - P_F) \sum_{k=0}^{\infty} k P_F^k = T + \frac{T_o P_F}{1 - P_F}$$

In the specific case of the numerical values here considered:

$$\bar{\vartheta} = 2.5 + \frac{3 \cdot 0.1}{0.9} = 2.833 \text{ ms}$$

Therefore, to guarantee that:

$$\rho = \lambda\bar{\vartheta} < 1$$

---

[5]The following formulas are useful to solve this problem:

$$\sum_{n=0}^{\infty} nx^n = \frac{x}{(1 - x)^2} \quad 0 \le x < 1 \qquad \sum_{n=0}^{\infty} n^2 x^n = \frac{x(1 + x)}{(1 - x)^3} \quad 0 \le x < 1$$

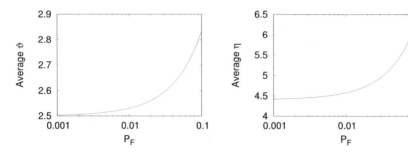

Fig. 5.19 Average service time and average waiting time for the stop-and-wait protocol as a function of the frame error probability P_F between 10^{-3} and 10^{-1}

we must impose that:

$$\lambda < \lambda_M = 352.9 \text{ frames/s}$$

Assuming then $\lambda = 0.8\,\lambda_M = 282$ frames/s and, therefore, $\rho = 0.8$, the system can be studied as an $\mathcal{M}/\mathcal{G}/1$ queue, where:

$$\bar{\eta} = \frac{\lambda\,E[\vartheta^2]}{2(1-\rho)}$$

We can calculate the mean square of the service time as:

$$E[\vartheta^2] = \sum_{k=0}^{\infty} \vartheta_k^2\, p_k = \sum_{k=0}^{\infty} (T + kT_o)^2\, P_F^k (1 - P_F) = (1 - P_F) \sum_{k=0}^{\infty} (T^2 + k^2 T_o^2 + 2kTT_o) P_F^k$$

$$= (1 - P_F) \left[\frac{T^2}{1 - P_F} + T_o^2 \frac{P_F(1 + P_F)}{(1 - P_F)^3} + \frac{2TT_o P_F}{(1 - P_F)^2} \right] = 9.14 \cdot 10^{-6}\ \text{s}^2$$

and then obtain:

$$\bar{\eta} = 6.44 \text{ ms}$$

The behavior of the average service time and average waiting time for the stop-and-wait protocol as a function of the frame error probability P_F between 10^{-3} and 10^{-1} is shown in Fig. 5.19.

5.4.3.2 Packet Payload Padding

A microcontroller gets messages from a set of sensors. The messages are queued inside the microcontroller RAM and are then sent to a monitoring system for further processing via a serial line with bit rate $C = 64$ kbits/s. The number and variety

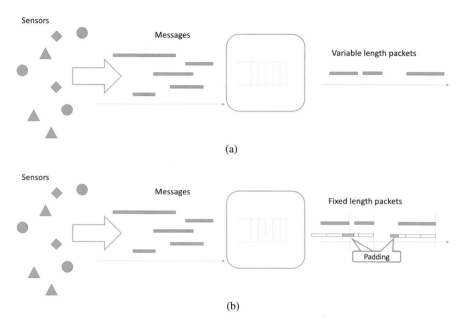

Fig. 5.20 Graphical example of the two options with variable-size packets (**a**) and fixed-size packets with padding (**b**)

of sensors is large enough to assume that messages arrive according to a Poisson process with average arrival rate $\lambda = 10$ messages/s. The messages are of random size, assumed independent and identically distributed according to an exponential distribution, with average $D = 512$ bytes.

The packet-based data transfer protocol between the microcontroller and the monitoring system has to be designed choosing among two possible alternatives, as illustrated in Fig. 5.20:

- Option 1: packets of variable size that encapsulate one single message as it comes from the sensors.
- Option 2: packets of fixed size equal to $D_0 = 200$ bytes, so that the original messages from the sensors must be split into segments fitting into the fixed-size packets. When a packet is not filled up completely, then dummy bytes are added to the payload, as many as required to reach the size D_0 (in packet switching this operation is usually called *padding*).

The overhead required to add a header to the each packet, including addressing and link control information between the microcontroller and the monitoring system, can be neglected for the purpose of this example.

Assuming an infinite queuing space at the output interface of the microcontroller, we want to find:

1. the traffic A_0 offered to the queuing system when option 1 is adopted;
2. the average packet waiting time $\bar{\eta}$ when option 1 is adopted;
3. the average number $\bar{N}$ of fixed-size packets per message when option 2 is adopted;
4. the average size L' of a message measured considering the padding, if needed, when option 2 is adopted;
5. the overhead due to padding, defined as $\mathcal{O} = \frac{L'-D}{D}$;
6. the traffic A'_0 offered to the queuing system when option 2 is adopted;
7. the average packet waiting time $\bar{\eta}'$ when option 2 is adopted.

Let us start considering option 1 and find the average service time when each message corresponds to a single variable-size packet:

$$\bar{\vartheta} = \frac{1}{\mu} = \frac{D}{C} = \frac{512 \cdot 8}{64000} = 64 \text{ ms}$$

from which we get:

$$A_0 = \rho = \lambda\bar{\vartheta} = 10 \cdot 64 \cdot 10^{-3} = 0.64 \text{ E}$$

and then, applying the formula for the $\mathcal{M}/\mathcal{M}/1$ system:

$$\bar{\eta} = \bar{\vartheta}\frac{\rho}{1-\rho} = 64 \cdot 10^{-3}\frac{0.64}{0.36} = 114 \text{ ms}$$

Now let us consider option 2. Indeed this case is a bit more complicated. We start by calculating the service time of a fixed-size packet:

$$\vartheta_s = \frac{D_0}{C} = \frac{200 \cdot 8}{64000} = 25 \text{ ms}$$

Finding the number of fixed-size packets per message is a problem similar to finding the cost of calls using tokens, as discussed in Sect. 2.6.1. The probability that a message fits within a single fixed-size packet can be written as the probability that the exponential service time ϑ under option 1 is less than or equal to the fixed service time ϑ_s under option 2, i.e.:

$$p_1 = \Pr\{\vartheta \leq \vartheta_s\} = F_\vartheta(\vartheta_s) = 1 - e^{-\mu\vartheta_s}$$

Similarly, the probability that a message does not fit within a single fixed-size packet but fits within two packets can be written as:

$$p_2 = \Pr\{\vartheta_s < \vartheta \leq 2\vartheta_s\} = F_\vartheta(2\vartheta_s) - F_\vartheta(\vartheta_s) = e^{-\mu\vartheta_s} - e^{-2\mu\vartheta_s}$$

In general, the probability that a message requires k fixed-size packet to be transmitted is:

$$p_k = \Pr\{(k-1)\vartheta_s < \vartheta \le k\vartheta_s\} = F_\vartheta(k\vartheta_s) - F_\vartheta((k-1)\vartheta_s)$$

$$= e^{-(k-1)\mu\vartheta_s} - e^{-k\mu\vartheta_s} = e^{-(k-1)\mu\vartheta_s}\left(1 - e^{-\mu\vartheta_s}\right) \quad \forall k = 1, 2, \ldots$$

Therefore, the average number of fixed-size packets can be computed as:

$$\bar{N} = \sum_{k=1}^{\infty} kp_k = \left(1 - e^{-\mu\vartheta_s}\right)\sum_{k=1}^{\infty} ke^{-(k-1)\mu\vartheta_s} = \frac{1}{1 - e^{-\mu\vartheta_s}} = \frac{1}{1 - e^{-25/64}} = 3.09$$

from which we can calculate the average size of a message, including the padding, as:

$$L' = \bar{N}D_0 = 618 \text{ bytes}$$

and the overhead as:

$$\mathcal{O} = \frac{L' - D}{D} = \frac{106}{512} = 0.21$$

The padding bits make messages appear longer than they are, and as a consequence the offered traffic increases. Neglecting correlations between packets generated from the same message, we can approximate the fixed-size packet arrival process as a Poisson process with arrival rate $\lambda' = \bar{N}\lambda = 30.9$ packets/s. Therefore:

$$A'_0 = \rho' = \lambda'\vartheta_s = 30.9 \cdot 25 \cdot 10^{-3} = 0.77 \text{ E}$$

Now, if we apply the formula for the $\mathcal{M}/\mathcal{D}/1$ system we get:

$$\bar{\eta}' = \frac{\vartheta_s}{2}\frac{\rho'}{1 - \rho'} = 41.85 \text{ ms}$$

From these results we conclude that queuing with fixed-size packets, under the conditions considered above, provides some advantages in terms of waiting time, even if the padding makes messages look longer and generates some additional traffic on the link.

5.4.3.3 Non-preemptive Priority Scheduling with Two Classes

Let us consider an $\mathcal{M}/\mathcal{G}/1$ queue with non-preemptive scheduling and two priority classes (class 1 with higher priority than class 2). Packets of the two classes have the same service time distribution, with the same average value $\bar{\vartheta} = 1$ s and the same mean square value $E[\vartheta^2]$. Let us assume that the total load ρ is split between traffic classes 1 and 2 as follows:

$$\rho_1 = x\rho$$

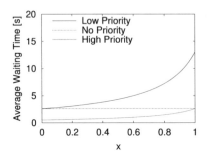 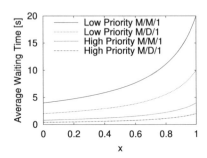

Fig. 5.21 $\mathcal{M}/\mathcal{G}/1$ with two priority classes and $\rho = 0.8$: average waiting time for the two classes as a function of the percentage x of high priority traffic. Left: case with $E[\vartheta^2] = 1.3\,\mathrm{s}^2$ and comparison with the case without priority. Right: cases with exponential and deterministic service time

$$\rho_2 = (1 - x)\rho$$

where $0 < x < 1$. In other words, changing the parameter x means changing the percentage of high priority traffic. We intend to analyze the behavior of the average waiting time of the two classes when x changes, for $\rho = 0.8$.

Using Cobham's formula (5.42) we obtain:

$$\bar{\eta}_1 = \frac{\bar{\vartheta}_P}{1 - \rho_1}$$

and:

$$\bar{\eta}_2 = \frac{\bar{\vartheta}_P}{(1 - \rho_1)(1 - \rho_1 - \rho_2)}$$

with:

$$\bar{\vartheta}_P = (\rho_1 + \rho_2)\frac{E[\vartheta^2]}{2\bar{\vartheta}} = \rho\frac{E[\vartheta^2]}{2\bar{\vartheta}} = 0.4\,E[\vartheta^2]$$

Therefore:

$$\bar{\eta}_1 = \frac{0.4\,E[\vartheta^2]}{1 - 0.8x}$$

$$\bar{\eta}_2 = \frac{0.4\,E[\vartheta^2]}{(1 - 0.8x)(1 - 0.8)}$$

Figure 5.21 shows the behavior of $\bar{\eta}_1$ and $\bar{\eta}_2$ as a function of x. In the left-hand graph the average waiting times are plotted in the case when $E[\vartheta^2] = 1.3\,\mathrm{s}^2$, and they are compared with the average waiting time of a generic packet when no priority scheduling is applied, under the same conditions (i.e., $\bar{\eta} = 2.6$ s). In the right-hand graph the two class waiting times are shown in case of an $\mathcal{M}/\mathcal{M}/1$ system ($E[\vartheta^2] = 2$ s^2) and an $\mathcal{M}/\mathcal{D}/1$ system ($E[\vartheta^2] = 1$ s^2), both with priority scheduling.

Obviously, when $x = 0$ the average waiting time of the low priority class converges to the value of a FIFO queue without priority and same load. Similarly for the high priority class when $x = 1$. Moreover, the graphs show an interesting behavior:

1. when x is small, $\bar{\eta}_1$ is significantly smaller than $\bar{\eta}_2$, while the latter is not too higher than the value of a queue without priority;
2. when x gets larger, then $\bar{\eta}_1$ is not much better than the value of a queue without priority, whereas $\bar{\eta}_2$ becomes much worse.

This last observation carries an important message: the priority scheduling works when the high priority traffic is only a little percentage of the total. Otherwise the only result is that low priority traffic is strongly penalized without significant improvement to high priority one.

5.4.3.4   Priority Scheduling for Multimedia Traffic

The traffic towards one of the output interfaces of a router can be classified into two categories, multimedia traffic from audio-video conferencing tools and normal traffic from typical office activities.

The multimedia traffic is approximately one third of the total traffic and is characterized by packets of fixed size equal to $D_m = 64$ bytes, with an average arrival rate of $\lambda_m = 1000$ packets/s. The remaining traffic is normal traffic characterized by variable length packets, with average $D_n = 512$ bytes and exponential distribution, and with an average arrival rate λ_n. The arrivals of both traffic classes can be considered to follow a Poisson process.

At first, let us assume that the router forwards all packets as a single traffic flow without any quality of service differentiation policy, with a queue to buffer packets that can be assumed infinite in size. We must evaluate:

- the total traffic offered to the output interface;
- the average service time $\bar{\vartheta}$ of a generic packet, assuming that the output interface of the router has capacity $C = 2.048$ Mbit/s;
- the average waiting time $\bar{\eta}$.

Then, we assume that after a proper software update the router is capable of implementing a QoS differentiation policy based on a non-preemptive priority scheduling mechanism, where multimedia traffic has the highest priority. We would like to evaluate:

- the average waiting time $\bar{\eta}_m$ for multimedia traffic;
- the average waiting time $\bar{\eta}_n$ for normal traffic.

To find the average service time of a generic packet, we must first compute the service time for multimedia traffic and the average service time for normal traffic, i.e.:

$$\vartheta_m = \frac{1}{\mu_m} = \frac{D_m}{C} = \frac{64 \cdot 8}{2.048 \cdot 10^6} = 0.25 \text{ ms}$$

$$\bar{\vartheta}_n = \frac{1}{\mu_n} = \frac{D_n}{C} = \frac{512 \cdot 8}{2.048 \cdot 10^6} = 2 \text{ ms}$$

The proportion between multimedia and normal traffic is given in terms of offered load. Therefore, we can write:

$$\lambda_m \vartheta_m = \rho_m = \frac{\rho}{3} \qquad \lambda_n \bar{\vartheta}_n = \rho_n = \frac{2}{3}\rho$$

where the total traffic offered to the output interface is:

$$\rho = \lambda \bar{\vartheta} = (\lambda_m + \lambda_n)\bar{\vartheta} = \lambda_m \vartheta_m + \lambda_n \bar{\vartheta}_n$$

From the known variables we can compute:

$$\rho_m = 1000 \cdot 0.25 \cdot 10^{-3} = 0.25 \text{ E}$$

$$\rho = 3 \rho_m = 0.75 \text{ E}$$

$$\rho_n = \frac{2}{3} \cdot 0.75 = 0.5 \text{ E}$$

$$\lambda_n = \frac{\rho_n}{\bar{\vartheta}_n} = \frac{0.5}{2 \cdot 10^{-3}} = 250 \text{ packets/s}$$

We can then compute the average service time of packets from any class by weighting the single class average service time using the related percentage of arrivals, i.e.:

$$\frac{\lambda_m}{\lambda} = \frac{1000}{1250} = 0.8 \qquad \frac{\lambda_n}{\lambda} = \frac{250}{1250} = 0.2$$

obtaining:

$$\bar{\vartheta} = \frac{\rho}{\lambda} = \frac{\lambda_m}{\lambda}\vartheta_m + \frac{\lambda_n}{\lambda}\bar{\vartheta}_n = 0.8 \cdot 0.25 \cdot 10^{-3} + 0.2 \cdot 2 \cdot 10^{-3} = 0.6 \text{ ms}$$

The average residual service time of a generic packet is given by the weighted average of the residual service time of the two types of packets. For multimedia traffic with packets of fixed size, the average residual service time is $\bar{\zeta}_m = \frac{\vartheta_m}{2}$, whereas for normal traffic, thanks to the memoryless property of the exponential distribution, it is $\bar{\zeta}_n = \bar{\vartheta}_n$. Considering that the probability of finding the server idle is $P_0 = 1 - \rho$, in which case the residual service time is zero, we can write:

$$(1 - \rho) \cdot 0 + \rho \bar{\zeta} = \rho_m \bar{\zeta}_m + \rho_n \bar{\zeta}_n$$

and then:

$$\bar{\zeta} = \frac{\rho_m}{\rho} \bar{\zeta}_m + \frac{\rho_n}{\rho} \bar{\zeta}_n = \frac{1}{3} \cdot \frac{0.25}{2} \cdot 10^{-3} + \frac{2}{3} \cdot 2 \cdot 10^{-3} = 1.375 \text{ ms}$$

The same result can be obtained by applying the general formula of the average residual service time (2.53), i.e.:

$$\bar{\zeta} = \frac{E[\vartheta^2]}{2\bar{\vartheta}} = \frac{1}{2\bar{\vartheta}} \left(\frac{\lambda_m}{\lambda} E[\vartheta_m^2] + \frac{\lambda_n}{\lambda} E[\vartheta_n^2] \right)$$

Considering that:

$$E[\vartheta_m^2] = \vartheta_m^2$$

and:

$$E[\vartheta_n^2] = 2\bar{\vartheta}_n^2$$

we obtain:

$$\bar{\zeta} = \frac{0.8 \cdot 0.25^2 \cdot 10^{-6} + 0.2 \cdot 2 \cdot 2^2 \cdot 10^{-6}}{2 \cdot 0.6 \cdot 10^{-3}} = 1.375 \text{ ms}$$

Finally, the average waiting time can be calculated from Eq. (5.34) as:

$$\bar{\eta} = \bar{\zeta} \frac{\rho}{1 - \rho} = 1.375 \cdot \frac{0.75}{0.25} = 4.125 \text{ ms}$$

When the router in capable of implementing non-preemptive priority scheduling, giving precedence to multimedia traffic, we must apply Cobham's formula (5.42). First we find that:

$$\bar{\vartheta}_p = \frac{1}{2} \left(\lambda_m \vartheta_m^2 + \lambda_n 2\bar{\vartheta}_n^2 \right) = \frac{1000 \cdot 0.25^2 \cdot 10^{-6} + 250 \cdot 2 \cdot 2^2 \cdot 10^{-6}}{2} = 1.03125 \text{ ms}$$

which can be also obtained considering that the weighted average residual service time is:

$$(1 - \rho) \cdot 0 + \rho \bar{\zeta} = 0.75 \cdot 1.375 \cdot 10^{-3} = 1.03125 \text{ ms}$$

And finally we have:

$$\bar{\eta}_m = \frac{\bar{\vartheta}_p}{1 - \rho_m} = 1.375 \text{ ms}$$

$$\bar{\eta}_n = \frac{\bar{\vartheta}_p}{(1 - \rho_m)(1 - \rho_m - \rho_n)} = 5.5 \text{ ms}$$

As expected, $\bar{\eta}_m$ decreases significantly (its value is one third of the waiting time without priority scheduling) and at the same time $\bar{\eta}_n$ experiences some increase (by one third of the value without priority scheduling). We can finally verify that:

$$\bar{\eta} = \frac{\rho_m}{\rho} \bar{\eta}_m + \frac{\rho_n}{\rho} \bar{\eta}_n = 4.125 \text{ ms}$$

5.4.3.5 Data, Voice, and Video Traffic with Priority

A router forwards data traffic towards a given destination through one of its interfaces, applying a non-preemptive priority scheduling among three traffic classes:

- **class 1**: video traffic with fixed-size packets of $D_1 = 1000$ bytes;
- **class 2**: voice traffic with fixed-size packets of $D_2 = 200$ bytes;
- **class 3**: data traffic with variable-size packets following a bi-modal distribution, such that 60% of the packets have size $D_{3,1} = 50$ bytes and the remaining 40% have size $D_{3,2} = 1500$ bytes.

The analysis performed on the global traffic exchanged by the router reported that 60% of the total traffic is data, 30% is voice and 10% is video.

Packet scheduling is configured such that class 1 has priority over class 2 and 3, and class 2 has priority over class 3. The goal is to find the average waiting time of the three classes when the interface link bit rate is $C = 1$ Mbit/s, the total offered traffic is $A_0 = \rho = 0.8$ E, and the length of the output queue at the interface can be considered infinite. In addition, we wish to compare the average waiting time of each class with the average waiting time in case no priority scheduling is applied and a simple FIFO scheduling is adopted.

To solve the problem we can apply the results obtained for the $\mathcal{M}/\mathcal{G}/1///\textbf{PRIO}$ queue, with $R = 3$ priority classes. The average waiting time per class can be found using Cobham's formula:

$$\bar{\eta}_r = \frac{\bar{\vartheta}_P}{(1 - S_{r-1})(1 - S_r)}$$

where

$$\bar{\vartheta}_P = \frac{1}{2} \sum_{i=1}^{R} \lambda_i E[\vartheta_i^2] \quad \text{e} \quad S_r = \sum_{i=1}^{r} \rho_i$$

At first, let us calculate $\bar{\vartheta}$ and $E[\vartheta^2]$ for the three traffic classes:

$$\bar{\vartheta}_1 = \frac{D_1 \times 8}{C} = 8 \text{ ms} \qquad E[\vartheta_1^2] = \bar{\vartheta}_1^2 = 64 \cdot 10^{-6} \text{ s}^2$$

$$\bar{\vartheta}_2 = \frac{D_2 \times 8}{C} = 1.6 \text{ ms} \qquad E[\vartheta_2^2] = \bar{\vartheta}_2^2 = 2.56 \cdot 10^{-6} \text{ s}^2$$

$$\bar{\vartheta}_3 = \frac{(0.6 \cdot D_{3,1} + 0.4 \cdot D_{3,2}) \times 8}{C} = 5.04 \text{ ms}$$

$$E[\vartheta_3^2] = 0.6 \cdot \left(\frac{D_{3,1} \times 8}{C}\right)^2 + 0.4 \cdot \left(\frac{D_{3,2} \times 8}{C}\right)^2 = 57.7 \cdot 10^{-6} \text{ s}^2$$

Knowing the percentage of each traffic class out of the total arrival rate, we can obtain the overall average service time as:

$$\bar{\vartheta} = \frac{\lambda_1}{\lambda}\bar{\vartheta}_1 + \frac{\lambda_2}{\lambda}\bar{\vartheta}_2 + \frac{\lambda_3}{\lambda}\bar{\vartheta}_3 = 4.3 \text{ ms}$$

Now we can calculate the load for each service class:

$$\rho_1 = \lambda_1 \bar{\vartheta}_1 = 0.1 \lambda \bar{\vartheta}_1 = 0.1 \rho \frac{\bar{\vartheta}_1}{\bar{\vartheta}} = 0.15$$

$$\rho_2 = \lambda_2 \bar{\vartheta}_2 = 0.3 \lambda \bar{\vartheta}_2 = 0.3 \rho \frac{\bar{\vartheta}_2}{\bar{\vartheta}} = 0.09$$

$$\rho_3 = \lambda_3 \bar{\vartheta}_3 = 0.6 \lambda \bar{\vartheta}_3 = 0.6 \rho \frac{\bar{\vartheta}_3}{\bar{\vartheta}} = 0.56$$

and then the arrival rate per class:

$$\lambda_1 = \frac{\rho_1}{\bar{\vartheta}_1} = 18.6 \text{ packets/s} \quad \lambda_2 = \frac{\rho_2}{\bar{\vartheta}_2} = 55.8 \text{ packets/s} \quad \lambda_3 = \frac{\rho_3}{\bar{\vartheta}_3} = 111.5 \text{ packets/s}$$

obtaining:

$$\bar{\vartheta}_P = 3.9 \text{ ms}$$

Now we know everything needed to apply Cobham's formula to each class:

$$\bar{\eta}_1 = \frac{\bar{\vartheta}_P}{1 - \rho_1} = 4.6 \text{ ms}$$

$$\bar{\eta}_2 = \frac{\bar{\vartheta}_P}{(1 - \rho_1)(1 - \rho_1 - \rho_2)} = 6 \text{ ms}$$

$$\bar{\eta}_3 = \frac{\bar{\vartheta}_P}{(1 - \rho_1 - \rho_2)(1 - \rho)} = 25.5 \text{ ms}$$

If we assume a simple FIFO scheduling, then the system can be studied as an $\mathcal{M}/\mathcal{G}/1$ system with load $\rho = 0.8$, average service time $\bar{\vartheta} = 4.3$ ms, and packet arrival rate $\lambda = \rho/\bar{\vartheta} = 185.9$ packets/s. In this case the average waiting time for any packet of any class is:

$$\bar{\eta} = \frac{\lambda \, E[\vartheta^2]}{2(1 - \rho)} = \frac{\lambda_1 \, E[\vartheta_1^2] + \lambda_2 \, E[\vartheta_2^2] + \lambda_3 \, E[\vartheta_3^2]}{2(1 - \rho)} = \frac{\bar{\vartheta}_P}{1 - \rho} = 19.4 \text{ ms}$$

If we compare this value to those calculated previously, we can see that traffic classes 1 and 2 actually have a performance improvement from priority scheduling, with a reduction in the average waiting time of 76.5% and 30.8%, respectively. On the other hand, low priority data traffic experiences an increase in the average waiting time of 31.2%.

As a final note, we can verify how the average unfinished workload $\bar{U}$ does not depend on the scheduling policy. In the $\mathcal{M}/\mathcal{G}/1///$**PRIO** case we have:

$$\bar{U} = \bar{\vartheta}_P + \rho_1 \bar{\eta}_1 + \rho_2 \bar{\eta}_2 + \rho_3 \bar{\eta}_3 = 19.4 \text{ ms}$$

which is equal to the average waiting time, and thus the average unfinished workload, in the $\mathcal{M}/\mathcal{G}/1$ case. In terms of Kleinrock conservation law, the invariant is:

$$\rho_1 \bar{\eta}_1 + \rho_2 \bar{\eta}_2 + \rho_3 \bar{\eta}_3 = \rho \bar{\eta} = 15.5 \text{ ms}$$

5.4.3.6   Token Bucket Scheduling

A packet switching node forwards to a given interface λ packets/s on average. Let us assume that the packets arrive according to a Poisson process. The packets are queued before being transmitted and are allowed to reach the server only when they obtain a specific permit, called *token*, from the queue scheduler. The tokens are generated with rate μ tokens/s with a deterministic inter-token interval $\vartheta_t = \frac{1}{\mu}$. The interface works as follows:

- when tokens are generated, they are stored in a dedicated memory at the interface, also called *bucket*;
- whenever a packet is served, it consumes a token from the bucket;
- if there are no tokens available, i.e., the bucket is empty, the first packet in the queue has to wait until a new token is generated;
- if there are no packets in the queue, the generated tokens are stored in the bucket up to a maximum of T; when the bucket is full, new tokens are discarded.

We are interested in characterizing the average waiting time for a packet before it can be transmitted. The queue we have to consider alternates between busy and idle periods, defined, respectively, as periods when there are packets being transmitted, and periods when there are no packets waiting in the queue. When a new packet arrives at the queue, there are two possibilities:

- the system is in an idle period, therefore the queue is empty and the packet is either transmitted immediately, if there is at least one token in the bucket, or is transmitted as soon as the next token is generated;
- the system is in a busy period, therefore the incoming packet joins the queue and must wait to obtain a token after all the previously queued packets obtain their respective tokens.

This system looks very similar to an $\mathcal{M}/\mathcal{D}/1$ queue, where the deterministic service time is represented by the inter-token interval. There is, however, a difference: when a new packet finds the server idle, as discussed above, it may happen that the bucket is empty and the packet must wait for a token to be generated, even with an idle server.

We can model this system using the formulas we know for the $\mathcal{M}/\mathcal{D}/1$ queue, but with some additional fixes. We know that the average number of customers in the queue for an $\mathcal{M}/\mathcal{D}/1$ system is:

$$A_c(\mathcal{M}/\mathcal{D}/1) = A'_c = \frac{\rho^2}{2(1-\rho)}$$

In the token bucket case we expect the actual average number of packets waiting in the queue to be larger than A'_c, because additional packets may be queued when the first packet in the queue is still waiting for a token to be generated. On average, the time to wait for a new token to be generated is:

$$t_w = \frac{\vartheta_t}{2} = \frac{1}{2\mu}$$

and the average number of packets arriving in such interval is:

$$N_w = \lambda t_w$$

From this observation we can compute an upper and a lower bound for the number of packets in the queue A_c, when the token bucket scheduling is applied:

1. the lower bound is $A_{cl} = A'_c$, which means assuming that a token is always available when a packet arrives at the queue, i.e., an optimistic assumption;
2. the upper bound is $A_{cu} = A'_c + N_w$, which means assuming that a token is never available when a packet arrives at the queue, i.e., a pessimistic assumption.

The actual average number of packets in the queue will be $A_{cl} \leq A_c \leq A_{cu}$, from which the average waiting time can be obtained applying Little's Theorem.

Exercises

1. A single server queuing system with an infinite queue is subject to Poisson arrivals with frequency $\lambda = 40$ arrivals/s and service times exponentially distributed with average $\bar{\vartheta} = 20$ ms. Find:

   (a) the range of values, measured in arrivals/s, that λ can take so that the system operates under stability conditions;
   (b) the probability of finding the server not busy;
   (c) the average number of customers in the queue;
   (d) the average waiting time for any customer and the average waiting time for customers that are being queued;
   (e) the probability that a customer spends inside the system more than $\delta_0 = 40$ ms;
   (f) the average frequency with which the customers leave the system.

2. A statistical multiplexer has N inputs, 1 output and a queue with infinite capacity. The arrival process at each input is Poisson with arrival rate $\lambda = 25$ packets/s and the service time is exponential with average $\bar{\vartheta} = 2$ ms. Find:

   (a) the maximum number N of inputs such that the system operates under stability conditions;
   (b) the maximum number N' of inputs such that the average queuing delay is $\bar{\eta} \leq 5$ ms.
   (c) the maximum number N'' of inputs such that the probability that a queued packet waits for more than 25 ms is less than or equal to 0.01.

3. A company operates across two sites, named A and B. The LANs of the two sites are equipped with a router each, acting as gateway towards the other site. A VPN is established between the two routers with a guaranteed capacity $C = 2$ Mbit/s. The internal bit rate of the LANs is $C_L \gg C$, therefore we can neglect queuing phenomena for incoming traffic directed to the LANs. On the other hand, queuing occurs for outgoing traffic. It is possible to assume that the packets arrive at each router according to a Poisson process with arrival rate

$\lambda = 120$ packets/s. Also, packets have random lengths exponentially distributed with average $D = 800$ bytes.

(a) Discuss whether the performance of each router has to be studied per se or if they can be studied together.

(b) Find the average number of packets in the outgoing output queues of the routers.

(c) Find the probability π_r that a packets has to wait in the queue.

(d) Find the average waiting time for the packets that have to wait.

(e) Since the company foresees an inter-site traffic growth of 30% in the next two years, evaluate the degradation in terms of average waiting time for the packets that have to wait under the new traffic conditions.

(f) Find by how much the guaranteed capacity C of the VPN must be increased in order to obtain the same average waiting time for the packets that have to wait as it was before the traffic increase.

4. A company spread over two sites needs to dimension the data connection between the routers connecting the two LANs in the two locations. Since the traffic between the sites is considered to be essentially symmetrical, it is requested to perform the evaluation for only one of the two directions, the data obtained being considered valid also for the other. Data traffic forecasts estimate that packets arrive at the output interface of the routers according to a Poisson process with the following characteristics:

- average initial arrival rate $\lambda(0) = 200$ packets/s;
- exponentially increasing arrival rate $\lambda(t)$ with an annual increase of 40%.

Packets are assumed to have random lengths, approximated by an exponential distribution with average $D = 800$ bytes. The planning objective is to guarantee the required quality of service for a period of 4 years, with any upgrade to the link capacity to be carried out at the beginning of each year, if necessary. The quality of service requirement is to keep the probability of packets waiting in the queue for less than $\epsilon_0 = 10$ ms, considering only packets that are actually queued, above 90%.

(a) Calculate the average arrival rate $\lambda(t)$ at the beginning of each year of operation.

(b) Find the line capacity needed at the beginning of each year of operation in order to guarantee the quality requirement throughout the whole year.

(c) Taking into account that the interconnection line may be acquired from a connectivity service provider as a multiple of the basic capacity unit $C_0 = 2$ Mbit/s, determine what capacity is needed initially and when upgrades are required.

(d) Find the queuing probability at the beginning of each year, as well as at the end of each year immediately before any link upgrade.

5. A company spread over two sites needs to dimension the data connection between the routers connecting the two LANs in the two locations. Since

the traffic between the sites is considered to be essentially symmetrical, it is requested to perform the evaluation for only one of the two directions, the data obtained being considered valid also for the other. Data traffic forecasts estimate that packets arrive at the output interface of the routers according to a Poisson process with an average arrival rate $\lambda(0) = 3700$ packets/s. It is also assumed that packets have random length, approximated by an exponential distribution with average $D = 1250$ bytes. The interconnection line can be acquired from a connectivity service provider as a multiple of the base capacity unit $C_0 = 10$ Mbit/s.

(a) Find the minimum number N_{min} of basic capacity units C_0 required to guarantee the stability of the router queues.
(b) Based on the value of N_{min} obtained above, find the average waiting time in the queue for a generic packet, as well as for a packet that is actually queued.
(c) Determine again the value of N_{min} such that an additional quality requirement is met, i.e., that at least 90% of the packets that actually end up in the queue remain in the queue for no more than 2 ms.
(d) Based on the previous point, find the probability that a packet needs to wait in the queue, the average waiting time in the queue for a generic packet, and the average waiting time for a packet that is actually queued.

6. A router is equipped with an output interface with very limited queuing space available, only 4000 bytes. The output channel capacity on the interface is $C = 1$ Mbit/s, and the router receives packets to be forwarded to that interface according to a Poisson arrival process with rate $\lambda = 100$ packets/s. The packet size is exponentially distributed with average $D = 1000$ bytes. Find:

(a) the probability of a packet being dropped because the queue is full;
(b) the probability of a packet being queued;
(c) the average number of customers in the system;
(d) the average waiting time for any customer.

7. A statistical multiplexer has N inputs, 1 output and a queue that can store up to $L = 3$ packets. The arrival process at each input is Poisson with arrival rate $\lambda = 25$ packets/s and the service time is exponential with average $\bar{\vartheta} = 2$ ms.

(a) Find the packet loss probability assuming $N = 12$ inputs.
(b) Compare the loss probability found above with the probability of having 3 customers in the queue in a similar system with infinite queuing space.

8. A network node has to process packets arriving as a Poisson process with average arrival rate $\lambda = 250,000$ packets/s. The packet processing requires two steps that have identical and independent random processing times, which are exponentially distributed with the same average $\bar{\vartheta}_p = 1$ μs. The two processing steps must necessarily be sequential, since the latter needs information resulting from the former. The processing is performed by two processors, P_1 and P_2, that

work as follows. P_1 takes a packet from the queue and processes it. When this job is done, P_1 checks P_2 and:

- case 1: if P_2 is free, the packet is passed over to it for the second step processing;
- case 2: if P_2 is still busy, the packet is held by P_1 until P_2 becomes available.

(a) Find the average total service time $\bar{\vartheta}_1$ of a packet in case 1, its variance $\sigma_{\vartheta_1}^2$ and the probability p_1 that this case occurs.

(b) Find the average total service time $\bar{\vartheta}_2$ of a packet in case 2, its variance $\sigma_{\vartheta_2}^2$ and the probability p_2 that this case occurs.

(c) Find the average total service time $\bar{\vartheta}$ of a generic packet and its variance σ_{ϑ}^2.

(d) Find the average waiting time of a generic packet $\bar{\eta}$.

9. A network node has to process packets arriving as a Poisson process with average arrival rate $\lambda = 250{,}000$ packets/s. The packet processing requires two operations that can be executed in parallel, which have identical and independent exponential random processing times with the same average $\bar{\vartheta}_p = 1.5 \ \mu$s. The processing is performed by two processors, P_1 and P_2. The service of a packet ends when both processors finish their processing task.

(a) Find the average service time of a packet $\bar{\vartheta}$.

(b) Find the variance of the service time σ_{ϑ}^2.

(c) Find the average waiting time of a generic packet $\bar{\eta}$.

10. A network node has to process packets arriving as a Poisson process with average arrival rate $\lambda = 250{,}000$ packets/s. The packet processing requires two steps, to be performed strictly in sequence by two processors, P_1 and P_2. The processing time in P_1 is deterministic with value $\vartheta_{p_1} = 1 \ \mu$s, while the processing time in P_2 is exponentially distributed with average $\bar{\vartheta}_{p_2} = 2 \ \mu$s. A packet is taken from the queue by P_1 only when P_2 has completed its processing task for the previous packet.

(a) Find the total average service time of a packet $\bar{\vartheta}$.

(b) Find the variance of the service time σ_{ϑ}^2.

(c) Find the average waiting time of a generic packet $\bar{\eta}$.

11. Find the lower and upper bounds of the average waiting time in a queue that applies a token bucket scheduling, assuming that $\lambda = 60$ packets/s arrive according to a Poisson process and that tokens are generated according to a deterministic process with rate $\mu = 100$ tokens/s.

12. An enterprise router acts as output gateway for the LANs of the company towards the Internet, and must deal with a packet arrival process that can be considered approximately Poissonian with average arrival rate $\lambda = 200$ packets/s. The packets are of two different types:

- 80% of them are of type A with fixed length $D_A = 64$ bytes;
- 20% of them are of type B with random length distributed exponentially and average $D_B = 512$ bytes.

The goal is to choose the capacity C of the link connecting the router to the Internet such that the following QoS objectives are met:

- type A packets experience an average waiting time $\bar{\eta}_A \leq 50$ ms;
- type B packets experience an average waiting time $\bar{\eta}_B \leq 200$ ms.

Also, for technical reasons C must be specified as a multiple of the basic capacity unit $C_0 = 64$ kbit/s, so that $C = n\, C_0$.
The capacity dimensioning must be performed by comparing the following implementation scenarios, assuming that the output interfaces of the router have an infinite queue:

(a) two different interfaces, the former with capacity C_A dedicated to traffic type A, the latter with capacity C_B dedicated to traffic type B;
(b) one single interface with capacity C_T that deals with all the traffic at once;
(c) one single interface with capacity C_P that deals with all the traffic at once, but implementing a non-preemptive priority scheduling such that type A packets have the highest priority.

13. A router forwards packets coming from a process monitoring system to an interface that has access to a dedicated channel with capacity $C = 128$ kbit/s. The router queues up packets arriving when the interface is busy transmitting another packet in a very large queuing space, which can be considered infinite. Two types of packets arrive from the monitoring system, both generated according to a Poisson process and both with exponentially distributed random lengths, with average size $D = 64$ bytes. The average arrival frequencies of the packets of the two traffic classes are $\lambda_1 = 30$ packets/s and $\lambda_2 = 180$ packets/s. Type 1 packets must be transmitted with a higher priority than type 2 packets, but without interrupting their transmission (non-preemptive priority scheduling).

(a) Find the average residual transmission time of a generic packet that may be currently in service at the time another packet arrives.
(b) Find the average queuing delay experienced by type 1 and type 2 packets, and compare it with the average queuing delay of a generic packet in the case of FIFO queuing discipline without priority.
(c) Evaluate what would be the average queuing delay experienced by type 1 packets in the case of preemptive priority scheduling.

Appendix A
Brief Introduction to Markov Chains

In this Appendix the basic results about Markov Chains (MCs) are briefly summarized. For the scope of this book, a MC is a random process evolving in time that can be described by means of a discrete set of values (the set may be either finite or infinite). The process randomly changes value either according to a discrete time model (changes in value may happen only at pre-defined time instants) or to a continuous time model (changes in value may freely happen at any time instant).

In Sect. A.1, we will define the basic concepts about MCs and focus on a discrete time model. Then in Sect. A.2 we will extend the concept to a continuous time model.

A.1 Discrete Time Markov Chains

Let us start considering a discrete time random process. The process can be described with a random variable X whose value changes at discrete time instants. X_i represents the value of X at time instant i. When the sequence of values describing the process $X_1, X_2, \ldots$ is such that:

$$\Pr\{X_n = j \,|\, X_{n-1} = i_{n-1}, X_{n-2} = i_{n-2}, \ldots, X_1 = i_1\}$$
$$= \Pr\{X_n = j \,|\, X_{n-1} = i_{n-1}\} \tag{A.1}$$

then the process is called a MC.

In brief this means that the value taken by the process at time n does not depend on the whole history of the process, but only on the previous value taken at time $n - 1$.

© Springer Nature Switzerland AG 2023
F. Callegati et al., *Traffic Engineering*, Textbooks in Telecommunication Engineering, https://doi.org/10.1007/978-3-031-09589-4

The value X_n taken by the process at time n is usually called the *state* of the process. In the simplest cases this is a numerical value, but more generally it could by an array of numbers, a vector, etc.

A.1.1   Transition and Steady State Probabilities

The probability in Eq. (A.1) is called *transition probability* from state i to state j at time instant n and in the following we will write it as:

$$P_{ij}(n) = \Pr\{X_n = j \mid X_{n-1} = i\} \tag{A.2}$$

More generally we may also define a transition probability from time instant m to time instant n as:

$$P_{ij}(m,n) = \Pr\{X_n = j \mid X_m = i\} \tag{A.3}$$

If Eq. (A.2) is known for every i and j and for all the time instants, then the behavior of the random process is completely known from the statistical point of view. Now let us give some important definitions about MCs.

A MC is *time homogeneous* if the transition probability does not depend on time but only on the states involved in the transition:

$$P_{ij}(n) = P_{ij} \; \forall n \tag{A.4}$$

From now on we will focus exclusively on time homogeneous MCs. Such a MC may be described with a state diagram, such as the one shown in Fig. A.1 for a 3-state MC. The circles represent the states, the value inside the circle is the numeric value associated with the state, and the arrows represent the transition probabilities between any possible pair of states.

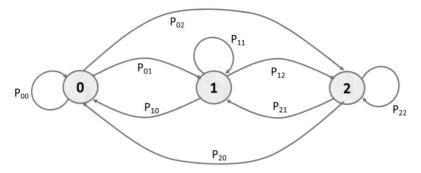

Fig. A.1 Example of state diagram for a discrete time MC

Considering all the P_{ij} $\forall i$ and $\forall j$, we can define the *transition probabilities matrix* as follows:

$$\mathscr{P} = \begin{bmatrix} P_{00} & P_{01} & P_{02} & \cdots \\ P_{10} & P_{11} & P_{12} & \cdots \\ P_{20} & P_{21} & P_{22} & \cdots \\ \vdots & \vdots & \vdots & \ddots \end{bmatrix} \tag{A.5}$$

In this matrix, the row index gives the starting state and the column index the landing state of the transition. If the number of states is finite and equal to N, then the matrix is a $N \times N$ square one, otherwise it is a matrix of infinite size.

Since P_{ij} is a probability, it follows that:

$$0 \le P_{ij} \le 1$$

$$\sum_{\forall j} P_{ij} = 1 \tag{A.6}$$

that is to say that the probability of landing in whatever state in the next transition is equal to 1, which is rather obvious intuitively. It follows that matrix $\mathscr{P}$ has two important properties:

- all the elements of the matrix take values between 0 and 1;
- each row is made by elements whose sum is 1.

We will see later that these properties have some important implications.

The *steady state probability* is the probability that the system is in state j at time n:

$$\pi_j(n) = \Pr\{X_n = j\} \tag{A.7}$$

The steady state probabilities can be grouped into a *steady state probability vector*:

$$\Pi(n) = \left[\pi_0(n), \pi_1(n), \ldots, \pi_j(n), \ldots\right] \tag{A.8}$$

A MC is *stationary* when the limit:

$$\lim_{n \to \infty} \pi_j(n) = \pi_j \quad \forall j \tag{A.9}$$

exists. That is to say that:

$$\lim_{n \to \infty} \Pi(n) = \Pi$$

The latter property guarantees that a limit for the steady state probability exists, but it does not say it is unique. In other words, the same MC might give in the limit different steady state probabilities, depending on the initial state.

A MC is *ergodic* when the limit in Eq. (A.9) exists and does not depend on the initial state $\Pi(0)$.

We can now link the steady state probabilities with the transition probabilities. Let us take a time homogeneous MC. If we know the steady state probability vector at time $n-1$ and the transition probabilities P_{ij}, then the steady state probability for state j at time n can be calculated as:

$$\pi_j(n) = \sum_{\forall i} \pi_i(n-1) P_{ij} \quad \forall j \tag{A.10}$$

This is obviously true for any state, and therefore Eq. (A.10) can be written in matrix form as:

$$\Pi(n) = \Pi(n-1)\mathscr{P} \tag{A.11}$$

A.1.2   Irreducible Markov Chains

Let us consider states i and j. We can say that they are *connected* if there is a non-zero probability to reach j from i in a finite number of time units.

> A MC is *irreducible* if all states are connected according to the previous definition. Otherwise the MC can be split in several MCs fully disjoint one another.

> State i is called *periodic* if the probability to start from i and be back in i is non-zero only for a given number of time units. In other words, state i is visited periodically.

> A MC is called *a-periodic* if it does not have any periodic state.

It is possible to prove that if a MC is a-periodic and irreducible, then the limit:

$$\lim_{n \to \infty} \pi_j(n) = \pi_j \quad \forall j$$

exists and does not depend on the initial state. In this case there are two possibilities:

1. $\pi_j = 0 \ \forall j$ and the MC does not have steady state probabilities,
2. $\pi_j > 0 \ \forall j$ and π_j is the steady state probability, which is unique whatever the initial state.

In the latter case, the MC is ergodic by definition and, after some time, the initial state does not have any influence on the statistical behavior of the MC, which depends only on the steady state probabilities and thus on the transition probabilities we use to calculate the steady state.

If we consider an ergodic MC, then when $n \to \infty$ Eq. (A.10) becomes:

$$\pi_j = \sum_{\forall i} \pi_i P_{ij} \quad \forall j \tag{A.12}$$

and Eq. (A.11):

$$\Pi = \Pi \mathscr{P} \tag{A.13}$$

The system expressed as in (A.13) is a homogeneous linear system of N equations in N unknowns (Π). Therefore the N unknowns are linearly dependent and there is not a unique solution to the system. It is possible to show that only one of these unknowns is dependent on the others, and therefore the solution depends on an arbitrary value.

We can fix this value recalling that the probabilities are conventionally defined with values between 0 and 1 and that the probability of the whole set of possible events is 1. Therefore:

$$\sum_{\forall j} \pi_j = 1 \tag{A.14}$$

As a matter of fact, with (A.14) we fix the values of the solution to Eq. (A.13). Equation (A.14) is not related to the specific behavior of the MC, but to the conventional definition about the values of probabilities and, in principle, could be different if we opted for a different conventional definition of the probability values (for instance, between 0 and 100 ad not between 0 and 1).

A.1.3   The Chapman–Kolmogorov Equations

The *flow* from state j to state i is defined as:

$$\phi_{ji} = \pi_j \cdot P_{ji} \tag{A.15}$$

The *outgoing* flow from state j is:

$$\phi_j^{\text{out}} = \sum_{\forall i} \pi_j \cdot P_{ji} = \sum_{\forall i} \phi_{ji} \tag{A.16}$$

and the *incoming* flow into state j is:

$$\phi_j^{\text{in}} = \sum_{\forall i} \pi_i \cdot P_{ij} = \sum_{\forall i} \phi_{ij} \tag{A.17}$$

Now we can write Eq. (A.12) as:

$$1 \cdot \pi_j = \sum_{\forall i} \pi_i P_{ij} \;\; \forall j$$

then using (A.6) (with i and j swapped):

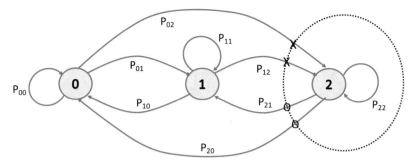

Fig. A.2 A graphical example of flow balancing

$$\sum_{\forall i} \pi_j P_{ji} = \sum_{\forall i} \pi_i P_{ij} \quad \forall j \tag{A.18}$$

and finally:

$$\phi_j^{\text{out}} = \phi_j^{\text{in}} \quad \forall j$$

which are also called *global balance equations* or *Chapman–Kolmogorov equations*.

The global balance equations say that the total flow outgoing from state j must be equal to the total flow incoming to state j, as shown in Fig. A.2. If we group several states into one big "macro-state" we can see that the global balance equations are valid through any closed line we can draw that cuts the state diagram of the MC.

A.1.4   The Markov Chain Behavior as a Function of Time

In the previous sections we have discussed the steady state behavior of the MC. Now let us discuss more in detail the transient or dynamic behavior. In particular, we are interested in finding the probability to be in state j at time n, given that at time 0 the MC was in state i.

Saying that the MC is in state i at time 0 means that $\pi_i(0) = 1$ and $\pi_j(0) = 0$ $\forall j \neq i$, i.e.:

$$\Pi(0) = \begin{bmatrix} 0 \cdots \underset{i}{1} \cdots 0 \end{bmatrix} \tag{A.19}$$

Now we can apply Eq. (A.11), getting:

$$\Pi(1) = \Pi(0)\mathscr{P}$$

$$\Pi(2) = \Pi(1)\mathscr{P} = \Pi(0)\mathscr{P}^2$$

$$\vdots$$ (A.20)

$$\Pi(n) = \Pi(n-1)\mathscr{P} = \Pi(0)\mathscr{P}^n$$

If the vector of the probabilities at time 0 $\Pi(0)$ is as in (A.19), when used in a product with $\mathscr{P}^n$ it selects the i-th row of $\mathscr{P}^n$.

If the MC is ergodic we know that the state probability will converge to the steady state values, i.e.:

$$\Pi(n) \to \Pi \quad \text{when} \quad n \to \infty \quad \forall \, \Pi(0)$$

Since this must be true whatever is the initial state, we can conclude that all the rows of the matrix $\mathscr{P}^n$, when $n \to \infty$, must be equal.

As a matter of fact, this can be an alternative methodology to solve the system in Eq. (A.13). Once we know $\mathscr{P}$, we can calculate:

$$\lim_{n\to\infty} \mathscr{P}^n$$

or approximate the limit with the product of $\mathscr{P}$ for itself computed n times, for a large value of n such that the difference between the values of the rows becomes negligible. From the computation point of view this may even be easier to implement than a conventional solution of the linear system.

A.1.5 Time Spent in a Given State

In the end let us analyze the time spent in a given state. Let us assume that the MC enters state j at time n_0. We would like to know the probability that the MC leaves state i for any other state at time n' such that $n' - n_0 = n$.

The probability that at a new time interval the MC stays in state j is P_{jj}, while $1 - P_{jj}$ is the probability of a state change. In a MC this is true for every time instant (recall (A.1)), since the transition depends only on the state at the previous time interval. Therefore we can write:

$$\Pr\{\text{the MC stays in } j \text{ for } n \text{ time intervals} \mid \text{just entered in } j\} = (1 - p_{jj})p_{jj}^{n-1}$$ (A.21)

This is the expression of the geometric probability mass function, which has the same memoryless properties as the exponential probability distribution. Therefore, it does not matter when the MC entered state j: if it is in state j now, we can calculate the probability that it will stay there for another n time intervals.

A.2 Continuous Time Markov Chain

The study of continuous time MC can be developed per se or as an extension of the previous results on discrete time MC. Let us start with the definition of a continuous time MC.

Given a continuous time random process $X(t)$, it is a *Markov Chain* if, for any sequence of time instants $t_1 < t_2 < \ldots < t_n$, we can write:

$$\Pr\{X(t_n) = j \mid X(t_{n-1}) = i_{n-1}, \ldots, X(t_1) = i_1\}$$
$$= \Pr\{X(t_n) = j \mid X(t_{n-1}) = i_{n-1}\} \tag{A.22}$$

The *transition probabilities* of the MC are:

$$P_{ij}(s, t) = \Pr\{X(t) = j \mid X(s) = i\} \quad t > s \tag{A.23}$$

and the *steady state probabilities* are:

$$\pi_j(t) = \Pr\{X(t) = j\} \tag{A.24}$$

We can define the *steady state probability vector* as:

$$\Pi(t) = \left[\pi_0(t), \pi_1(t), \ldots, \pi_j(t), \ldots\right]$$

A continuous time MC is *time homogeneous* if $P_{ij}(s, t)$ does not depend on a specific instant s, but only on the difference $t - s$.

A continuous time MC is *stationary* if the limit:

$$\lim_{t \to \infty} \Pi(t) = \Pi \tag{A.25}$$

exists. Moreover, if Π does not depend on the initial state from which the process started, then the MC is *ergodic*.

If a continuous time MC is a-periodic and irreducible, then the limit (A.25) always exists and what already discussed in Sect. A.1 applies.

We can now split the time axis in intervals of the same size Δt, emulating a discrete time process. Recalling Eq. (A.10) we get:

$$\pi_j (t + \Delta t) = \sum_{\forall i} \pi_i (t) \, P_{ij} (\Delta t) \quad \forall j \tag{A.26}$$

which can be written as:

$$\frac{\pi_j (t + \Delta t) - \pi_j (t)}{\Delta t} = \frac{\sum_{\forall i} \pi_i (t) \, P_{ij} (\Delta t) - \pi_j (t)}{\Delta t} \tag{A.27}$$

If we now compute the limit for $\Delta t \to 0$ we obtain:

$$\frac{d}{dt} \pi_j (t) = \lim_{\Delta t \to 0} \frac{\pi_j (t + \Delta t) - \pi_j (t)}{\Delta t}$$

$$= \lim_{\Delta t \to 0} \left[\frac{P_{jj} (\Delta t) - 1}{\Delta t} \pi_j (t) + \sum_{\forall i \neq j} \frac{P_{ij} (\Delta t)}{\Delta t} \pi_i (t) \right]$$

$$= \pi_j (t) \lim_{\Delta t \to 0} \frac{P_{jj} (\Delta t) - 1}{\Delta t} + \sum_{\forall i \neq j} \pi_i (t) \lim_{\Delta t \to 0} \frac{P_{ij} (\Delta t)}{\Delta t} \tag{A.28}$$

We now define the *transition frequencies* (or *rates*) as:

$$q_{ij} = \lim_{\Delta t \to 0} \frac{P_{ij} (\Delta t)}{\Delta t} \tag{A.29}$$

$$q_{jj} = \lim_{\Delta t \to 0} \frac{P_{jj} (\Delta t) - 1}{\Delta t} \tag{A.30}$$

The matrix:

$$\mathcal{Q} = \begin{bmatrix} q_{00} & q_{01} & q_{02} & \cdots \\ q_{10} & q_{11} & q_{12} & \cdots \\ q_{20} & q_{21} & q_{22} & \cdots \\ \vdots & \vdots & \vdots & \ddots \end{bmatrix} \tag{A.31}$$

is called the *transition rate matrix*. Similarly to the definition in (A.5), the row index i identifies the departing state while the column index j identifies to landing state. The corresponding value in the matrix is the transition rate from i to j.

Equation (A.28) can be rewritten as

$$\frac{d}{dt}\pi_j(t) = q_{jj}\pi_j(t) + \sum_{\forall i \neq j} q_{ij}\pi_i(t) = \sum_{\forall i} q_{ij}\pi_i(t) \tag{A.32}$$

and the whole set of Eqs. (A.32) can be written as:

$$\frac{d}{dt}\Pi(t) = \Pi(t)\,\mathcal{Q} \tag{A.33}$$

The coefficients q_{ij} have the dimension of a frequency, and that is the reason why they are called transition rates. If we compute $q_{ij}dt$, we get the probability to move from state i to state j during a very small time period dt.

In this case the rows of matrix $\mathcal{Q}$ are not probabilities and do not sum up to 1. They actually sum up to 0, given the way the transition rates have been defined:

$$\sum_{\forall j} q_{ij} = \lim_{\Delta t \to 0} \frac{1}{\Delta t}\left[\sum_{\forall i \neq j} P_{ij}(\Delta t) + P_{jj}(\Delta t) - 1\right] = 0 \tag{A.34}$$

If the MC is ergodic, then at some stage the time does not influence the behavior of the system anymore and therefore:

$$\frac{d\pi_j(t)}{dt} = 0 \tag{A.35}$$

so Eq. (A.33) becomes:

$$\Pi\,\mathcal{Q} = 0 \tag{A.36}$$

For a continuous time MC equation (A.36) is the same as Eq. (A.13) for a discrete time MC. For a continuous time MC we can also draw a state diagram were the transition rates take the place of the transition probabilities, as shown in Fig. A.3. Obviously in this case the rate to stay in the same state does not appear, because it has no physical meaning (i.e., we cannot measure the rate of a state transition that does not happen).

Again in analogy with the discrete time MC, we can define the flow from state j to state i (ϕ_{ji}), the *incoming flow* to state j (ϕ_j^{in}), and the *outgoing flow* from state j (ϕ_j^{out}) as in Eqs. (A.15), (A.16) and (A.17) by simply using the transition rates q_{ji} instead of the transition probabilities P_{ji}.

The global balance equations are given by the combination of Eqs. (A.34) and (A.36) and can be written as:

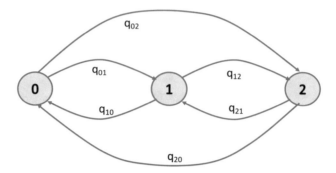

Fig. A.3 Example of state diagram for a continuous time MC

$$\sum_{\forall i} \pi_j q_{ji} = \sum_{\forall i} \pi_i q_{ij} \quad \forall j \tag{A.37}$$

For a continuous time MC the flow balance also applies:

$$\phi_j^{\text{out}} = \phi_j^{\text{in}} \quad \forall j$$

which is an alternative way to find Eq. (A.36).

   Given an ergodic continuous time MC, the steady state probabilities can be calculated by solving the system of equations that combines (A.36) and (A.14).

A.2.1   The Time Spent in a State

If τ_j is the time spent in state j, after entering it of course, it is possible to prove that:

1. τ_j is a random variable with an exponential distribution, i.e.:

$$\Pr\{\tau_j < t\} = 1 - e^{-q_j t} \tag{A.38}$$

   where

$$q_j = \sum_{\forall i \neq j} q_{ji} \tag{A.39}$$

2. when the process leaves state j it goes to state i with probability:

$$P_{ji} = \frac{q_{ji}}{q_j}$$

Appendix B
The Embedded Markov Chain for the $\mathcal{M}/\mathcal{G}/1$ System

The main quantities of interest for an $\mathcal{M}/\mathcal{G}/1$ system are the same as in an $\mathcal{M}/\mathcal{M}/1$ system. In addition, we define:

t_n	the time epoch in which the n-th customer leaves the system;
$k(t_n) = k_n$	the number of customers in the system after time epoch t_n;
a_n	the number of arrivals during the service time of the n-th customer;
a	the number of arrivals during a service time at the equilibrium (in this case there is no need to consider the time index);
α_i	the probability $\Pr\{a = i\}$ in a given service time at the equilibrium;
σ_{ϑ}^2	the variance of the service time.

When considering the system state described by means of k_n, i.e., the number of customers left in the system immediately after a departure, we have two possibilities.

- The customer departing at t_n leaves the system not empty: in this case, the $n + 1$-th customer, first in line waiting in the queue, begins its service, during which a_{n+1} new customers arrive; then it leaves at t_{n+1}.
- The customer departing at t_n leaves the system empty: in this case, when the $n+1$-th customer arrives, it begins its service immediately, during which a_{n+1} new customers arrive; then it leaves at t_{n+1}. There is one additional arrival between t_n and t_{n+1} in this case, with respect to the previous one.

Based on those two possibilities, we can write the equation that links the values of the state at two consecutive departure time instants, i.e., k_n and k_{n+1}:

© Springer Nature Switzerland AG 2023
F. Callegati et al., *Traffic Engineering*, Textbooks in Telecommunication
Engineering, https://doi.org/10.1007/978-3-031-09589-4

$$k_{n+1} = \begin{cases} k_n - 1 + a_{n+1} & k_n > 0 \\ a_{n+1} & k_n = 0 \end{cases} \tag{B.1}$$

If we define the unit step function as:

$$u(k_n) = \begin{cases} 1 & k_n > 0 \\ 0 & k_n = 0 \end{cases}$$

then Eq. (B.1) can be rewritten as:

$$k_{n+1} = k_n - u(k_n) + a_{n+1} \tag{B.2}$$

It is worth noting that the state at time t_{n+1} depends only on the state at time t_n and on the number of arrivals during the service of customer $n + 1$. Therefore, such a system is a Markov Chain, although not a BD process because transitions can happen between non-adjacent states. However, this does not mean that the $\mathcal{M}/\mathcal{G}/1$ queuing system can be studied as a Markov Chain, it means only that the system *observed at customer departure times* can be studied as a Markov Chain. This is usually called an *embedded* (or *hidden*) Markov Chain.

If $\rho < 1$, it is possible to prove that the embedded Markov Chain is ergodic, and therefore a steady state probability distribution $\Pi = \{\pi_0, \pi_1 \ldots \pi_k \ldots\}$ exists. We first study the $\mathcal{M}/\mathcal{G}/1$ embedded Markov Chain in Sect. B.1, and then show in Sect. B.2 that the results obtained at the customer departure times are actually valid at any time instant.

B.1   Steady State Probabilities of the Number of Customers in the System at Departure Times

We are interested in finding the state probabilities at the equilibrium describing the embedded Markov Chain:

$$\pi_k = \lim_{n \to \infty} \Pr\{k_n = k\}$$

To calculate π_k we need to know the transition probabilities of the Markov Chain, which are a function of the number of arrivals within a service time. For the case $k_n \neq 0$, using (B.2) we have:

$$P_{ij}(n, n + 1) = \Pr\{k_{n+1} = j \mid k_n = i\} = \Pr\{a_{n+1} = j - i + 1\}$$

At the equilibrium this leads to:

$$\lim_{n\to\infty} P_{ij}(n, n+1)$$

$$= \lim_{n\to\infty} \Pr\{a_{n+1} = j - i + 1\} = \Pr\{a = j - i + 1\}$$

$$= \alpha_{j-i+1} = \int_0^\infty \frac{(\lambda t)^{j-i+1}}{(j-i+1)!} e^{-\lambda t} f_\vartheta(t)dt \tag{B.3}$$

once the probability density function $f_\vartheta(t)$ of the service time is known. For the case $k_n = 0$, the transition probabilities are:

$$P_{0j}(n, n+1) = \Pr\{a_{n+1} = j\}$$

and at the equilibrium:

$$\lim_{n\to\infty} P_{0j}(n, n+1) = P_{0j} = \alpha_j$$

The problem of finding π_k is equivalent to solving the Markov Chain by means of the well known equation:

$$\Pi = \Pi \mathscr{P}$$

where Π is the vector of the state probabilities and $\mathscr{P}$ is the matrix of the transition probabilities. In this case $\mathscr{P}$ has the form:

$$\mathscr{P} = \begin{bmatrix} \alpha_0 & \alpha_1 & \alpha_2 & \alpha_3 & \cdots \\ \alpha_0 & \alpha_1 & \alpha_2 & \alpha_3 & \cdots \\ 0 & \alpha_0 & \alpha_1 & \alpha_2 & \cdots \\ 0 & 0 & \alpha_0 & \alpha_1 & \cdots \\ \vdots & \vdots & \vdots & \vdots & \ddots \end{bmatrix} \tag{B.4}$$

Therefore we can write:

$$\pi_j = \pi_0 \alpha_j + \sum_{i=1}^{j+1} \pi_i \alpha_{j-i+1} \quad j = 0, 1, 2, \ldots \tag{B.5}$$

This set of equations is not easy to solve directly. One synthetic expression can be found by means of the z-transform. Recall that the z-transform of a discrete series of numbers $\{x_i\}$ is defined as:

$$X(z) = \sum_{i=0}^\infty z^i x_i$$

Therefore we can use the z-transforms of the $\{\pi_j\}$ and $\{\alpha_j\}$ series:

$$\Pi(z) = \sum_{j=0}^{\infty} z^j \pi_j \qquad \alpha(z) = \sum_{j=0}^{\infty} z^j \alpha_j$$

Then, if we multiply equation (B.5) by z^j and sum for all values of j we obtain:

$$\Pi(z) = \sum_{j=0}^{\infty} \pi_j z^j = \sum_{j=0}^{\infty} \left[\pi_0 \alpha_j z^j + \frac{1}{z} \sum_{i=1}^{j+1} \pi_i \alpha_{j-i+1} z^{j+1} \right]$$

$$= \sum_{j=0}^{\infty} \left[\pi_0 \alpha_j z^j + \frac{1}{z} \sum_{i=0}^{j+1} \pi_i \alpha_{j-i+1} z^{j+1} - \frac{\pi_0 \alpha_{j+1} z^{j+1}}{z} \right]$$

$$= \pi_0 \alpha(z) + \frac{\alpha(z)\Pi(z)}{z} - \frac{\pi_0 \alpha(z)}{z} \tag{B.6}$$

This equation is obtained by writing:

$$\sum_{j=0}^{\infty} \sum_{i=0}^{j+1} \pi_i \alpha_{j-i+1} z^{j+1} = \sum_{j=1}^{\infty} \sum_{i=0}^{j} \pi_i \alpha_{j-i} z^j = \sum_{j=0}^{\infty} \sum_{i=0}^{j} \pi_i \alpha_{j-i} z^j - \pi_0 \alpha_0$$

and noting that:

$$\sum_{j=0}^{\infty} \sum_{i=0}^{j} \pi_i \alpha_{j-i} z^j$$

is the transform of the discrete convolution of the two sequences $\{\pi_j\}$ and $\{\alpha_j\}$, which is equal to the product of the z-transforms $\Pi(z)$ and $\alpha(z)$. Furthermore:

$$\sum_{j=0}^{\infty} \frac{\pi_0 \alpha_{j+1} z^{j+1}}{z} = \sum_{j=0}^{\infty} \frac{\pi_0 \alpha_j z^j}{z} - \frac{\pi_0 \alpha_0}{z} = \frac{\pi_0 \alpha(z)}{z} - \frac{\pi_0 \alpha_0}{z}$$

Solving equation (B.6) for $\Pi(z)$ and recalling that $\pi_0 = 1 - \rho$, we finally obtain the expression for the z-transform of the steady state probabilities:

$$\Pi(z) = \frac{(1 - \rho)(1 - z)\alpha(z)}{\alpha(z) - z} \tag{B.7}$$

This formula can be used to obtain π_k formally or numerically when $\alpha(z)$ is known.

B.2 Steady State Probabilities at Generic Time Instants

It is possible to prove that the state probabilities calculated at t_n are equal to the state probabilities at any other time instant. To do so we first focus our attention on the instants of arrivals and departures, and then proceed as follows.

We consider a (large) time interval T and define:

P_k^a the state probabilities at the arrival instants;

P_k^d the state probabilities at the departure instants, which we already know, at least in terms of their z-transforms;

$a_k(T)$ the number of arrivals in T that occur when there are k customers in the system, i.e., the number of events that bring the system from state k to state $k + 1$;

$d_k(T)$ the number of departures in T that occur when there are $k + 1$ customers in the system, i.e., the number of events that bring the system from state $k + 1$ to state k;

$a(T)$ the total number of arrivals in T;

$d(T)$ the total number of departures in T.

Since the state of the system changes by one unit at a time, we can write:

$$\mid a_k(T) - d_k(T) \mid \leq 1$$

In fact state k can be visited only after a birth event happening in state $k - 1$ or a death event happening in state $k + 1$. Therefore, the numbers of arrivals to and departures from state k in the period T are either equal or differ by one unit in the following cases:

- when the system starts from a state $p \leq k$ and ends in a state $q \geq k$, in which case $a_k(T) = d_k(T) + 1$;
- when the system starts from state $p \geq k$ and ends in a state $q \leq k$, in which case $a_k(T) = d_m(T) - 1$.

Furthermore, it is obvious that:

$$d(T) = a(T) + k(0) - k(T)$$

Now, the state probabilities at the instants of arrivals and departures can be approximated as:

$$P_k^a = \lim_{T \to \infty} \frac{a_k(T)}{a(T)} \quad \forall k$$

$$P_k^d = \lim_{T \to \infty} \frac{d_k(T)}{d(T)} \quad \forall k$$

and then we can write:

$$\lim_{T \to \infty} \frac{d_k(T)}{d(T)} = \lim_{T \to \infty} \frac{a_k(T) + d_k(T) - a_k(T)}{a(T) + k(0) - k(T)} = \lim_{T \to \infty} \frac{a_k(T)}{a(T)} \tag{B.8}$$

since $d_k(T) - a_k(T)$ is bounded by -1 and 1, and $k(T)$ and $k(0)$ are finite if the system is stable. This proves that:

$$P_k^a = P_k^d \tag{B.9}$$

but since the arrival process is independent of the state of the system and is memoryless, observing the system at the arrival instants is equivalent to observing it at random time instants. Therefore, the state probabilities at the arrival instants must be equal to the state probabilities at any time instant, and therefore they are equal to the state probabilities at the departure time instants, which were obtained for the embedded Markov Chain.

Index

A
Average customer arrival rate, 48
Average customer service time, 8

B
Bandwidth sharing, 125–129, 141
Birth-death process, 45–63
Birth rate, 49, 52
Blocking probability, 54, 56, 60, 61, 66, 67,
  70, 72–74, 76–81, 83, 86, 87, 108,
  117–120, 122, 123, 125–129, 131–139,
  153–155
Busy period, 143, 197

C
Call center dimensioning, 102–104, 109
Circuit switching, 65–115, 129–133
Complete partitioning, 125
Complete sharing, 125
Congestion, 46–48, 57, 66, 68–71, 74, 76, 78,
  91, 95, 113, 121, 146–150
Customer arrival time, 29
Customer departure time, 10, 11, 151

D
Death rate, 49, 52, 93

E
Erlang, 1, 2, 24, 26, 27, 39, 43, 55, 69, 76–79,
  82, 86, 89, 91, 93, 95, 110, 113, 131,
  133–135, 146, 155, 172, 174–176

Erlang B formula, 55, 70–71, 133
Erlang C formula, 95

F
First In, First Out (FIFO), 7, 8, 12, 92, 98–102,
  112, 143, 158, 170–172, 177, 191, 196,
  202
Fixed routing network, 66, 120–122

G
General service time, 170

I
Idle time, 143
Improving server utilization, 67, 72, 86, 87, 94,
  133, 155, 179
Infinite queue size, 155, 168, 185
Inter-arrival time, 19–20, 24, 41–43, 51, 53,
  152

K
Kendall's notation, 6, 7, 11

L
Last In, First Out (LIFO), 7
Little, 4
Little's Theorem, 3–4, 8–10, 97, 147, 179, 198
Lost traffic, 9, 112, 115

© Springer Nature Switzerland AG 2023
F. Callegati et al., *Traffic Engineering*, Textbooks in Telecommunication
Engineering, https://doi.org/10.1007/978-3-031-09589-4

Printed in the United States
by Baker & Taylor Publisher Services